Integrated Farming System Models

NIPA® GENX ELECTRONIC RESOURCES & SOLUTIONS P. LTD.
New Delhi-110 034

About the Editors

Dr. Adyant Kumar is working as an Assistant Professor cum Junior Scientist in the Department of Agronomy at Dr. Kalam Agricultural College, (Bihar Agricultural University), Kishanganj, Bihar. He completed his Doctorate and Master's degrees in Agriculture with a specialisation in Agronomy from BCKV, West Bengal and RAU, Pusa (Bihar), respectively. He is the recipient of the Fakhruddin Ali Ahmed Award for Outstanding Research in Tribal Farming Systems, 2021 given by ICAR under MoA&FW, GoI. He is also awarded with various awards & accolades viz. Best Oral Presentation, Best Thesis, Young Scientist and Young Achiever award as well as Valued Reviewer recognition from Springer Nature etc. He performed as the Co-PI in the project "Study of Millets production and consumption pattern in the last 20 years in Jharkhand" sanctioned by DRDM Tribal Welfare Research Institute, Ranchi. He had also contributed as a Mentor under the GOAL (Going Online as Leaders) Program for the year 2020-21 which was organised by Ministry of Tribal Affairs, Govt. of India in collaboration with Meta (Facebook). He is the Lifetime Member in Indian Society of Agronomy, and Crop and Weed Science Society as well as in the Board of Advisors of NAVADRISHTI, Ranchi, Jharkhand. He has guided six M.Sc. students during his teaching period. He had written about 22 Research articles, 01 Review articles, 02 Book, 01 Souvenir, 05 Book chapters and 11 Popular articles in English and regional language for academicians, researchers, students and farming communities.

Dr. Sudhanand Prasad Lal is currently an Assistant Professor cum Scientist, in the Department of Extension Education at PG College of Agriculture, RPCAU, Pusa, MoA&FW, Govt. of India. He completedhis B.Sc. Ag.(Hons.) from CSA University, Kanpur in 2012 and received Chancellor's gold medal. He obtained his MSc from, ICAR-National Dairy Research Institute, Karnal-132001, (Haryana) in 2014 & was awarded merit certificate in the Science & Management group. He completedhis Ph.D. in Agricultural Extension Education on 12th March 2018 and received Ph.D. best thesis Director's Gold medal at ICAR-NDRI, Karnal from Dr. Trilochan Mohapatra, Secretary (DARE) & Director General (ICAR) Krishi Bhavan, New Delhi. He is the recipient of NTS in B.Sc., JRF in M.Sc., and SRF in Ph.D. He is also the recipient of Certificate of Excellence Award from Elsevier Researcher Academy,

Amsterdam, Netherlands. He has written 40 Research articles,01 Review article, 02 Short communications,03 Books, 06 Book chapters and 20 popular articles. He was invited as an expert by 'The Times Higher Education Academic Reputation Survey in partnership with Elsevier for the 2019 Times Higher Education World Reputation Rankings and the 2019 World University Rankings, supported by UNICEF. Dr. Lal has received citations from DRDO, IIT-Delhi, CAB Direct and FAO.

Dr. Milind B. Katare, Ph.D. (Aquaculture) is presently working as a Subject Matter Specialist (Fisheries) at Krishi Vigyan Kendra, East Garo Hills-794111 Meghalaya under the administrative control of Central Agricultural University, Imphal, Manipur. He obtained his B.F.Sc. (Bachelor of Fishery Science) from the College of Fishery Science, Maharashtra Animal and Fishery Sciences University, Nagpur followed by an M.F.Sc. (Master of Fishery Science) in Aquaculture from the College of Fisheries, Mangalore under Karnataka Veterinary, Animal and Fisheries Sciences University, Bidar. He has completed the Doctorate (Ph.D.) in Aquaculture from ICAR-Central Institute of Fisheries Education, Mumbai, Maharashtra. His publications are includedin the library of FAO-AGRIS of the United Nations and ICAR, New Delhi, especially in the areas of fish endocrinology, physiology, genetics and breeding. He has been honoured with several awards viz., Young Scientist, Best KVK Scientist and Excellence in Extension Scientist Award etc. He isa member of the Board of Studies for Centurion University of Technology and Management,Odisha-761211, INDIA (Established Vide Orissa Act 4 of 2010). He is presently acting as a Valued Reviewer for esteemed Springer and Elsevier journals.

Dr. Reena Kumari has been working as Assistant Professor forthe pasteight years at Navsari Agricultural University in Navsari, Gujarat. She has done her doctorate and master degree in Agricultural Engineering with specialisation in soil and water conservation engineering from Banaras Hindu University, Varanasi. She is working on conservation agriculture and has recommended three technologies related to conservation agriculture for the farming community of southern Gujarat. She is member of many reputed societies and published over 25 research papers in national and international journals. She has written 12 book chapters and 5 popular articles in English and regional languages for academicians, researchers, students and farming communities. She has received Young Scientist award, Women Scientist award, Best Poster Presentation award by the different reputed societies. She is Principal Investigator and Co-Principal Investigator of different University Research Projects. She has participated and presentedher work in different international and national conference organised by reputed institutes and Societies. She has also co-organised 4 workshop cum training programmes for students, researchers, academicians and farmers

Dr. Pratibha Kumari, is presently working as a Subject Matter Specialist (Soil Science) at Krishi Vigyan Kendra, Araria (Bihar Agricultural University) Bihar. She obtained her Doctorate and Master's degree in soil science and agricultural chemistry with specialisation in soil fertility from Banaras Hindu University Varanasi. She has a good knowledge about the conservation agriculture for natural resource management. She is passionate to empower the farmers with the latest cutting-edge technology by transferring her research experience from the lab to the field. She worked at the International Rice Research Institute Bihar hub as a Specialist in Agriculture Research & Development for almost 5 years. During work at International Rice Research Institute (IRRI) as a team member worked to developed web-based platform (Crop Manager for Rice based System) for field-specific information on crop and nutrient management to increase yields and income forrice farmers.She has published more than 15 research papers in national and international journals. She has been awarded as young scientist and excellence in research award by reputed institutes and Societies of India. She has authored many book chapters, books and essential literature for academicians, researchers, scientists, students and the farming community in international, national and regional language. Dr. Pratibha is active on various platforms including Radio talk, Kisan Mela, Kisan Gostthi and Cooperative societiesto spread awareness and educate farmers about the latest trends and efficient farming practices. She shares her experience in trainings workshops and conferences organised by international and national Institute.

Dr. D.K. Jayswal is currently working as an Assistant Professor-cum- Junior Scientist, in the Department of Horticulture (Fruit & Fruit Technology), at Bihar Agricultural College, (Bihar Agricultural University), Bhagalpur, Bihar. He completed his Doctorate and Master's degree in Agriculture with a specialisation in Horticulture – Fruit Science from JNKVV, Jabalpur (M.P.) in 2017 and 2014 subsequently. He has significantly contributed to standardising the severity of pruning and developed a combination of organic, inorganic and bio-fertilizers as INM for improving the growth, yield and quality of guava fruit during his Ph.D. degree programme. He is a recipient of several awards and certificates in which some of explicit awards are Young Scientist Award by the Society of Tropical Agriculture, Dr. N.C. Gautam Best Ph.D. Thesis Award in Horticulture – 2017 by the Society for Horticultural Research and Development, Ghaziabad, Professional Excellence Award by the Indian Association of Veterinary Pathologists (Jharkhand Chapter), Research Excellence Award-2020 by the Institute of Scholars, Bangalore etc. He has also participated in cultural activities and won the 2nd Prize in Debate and 3rd Prize in One Act Play in Agri-unifest. He has published 18 Research and Review articles in NAAS rated peer-reviewed journals, 03 Book chapters, 06 popular articles and more than 10 abstracts in souvenirs. He is a Life Member of the Society for Horticultural Research and Development, The Indian Science Congress Association and the Indian Society for Arid Horticulture. He has good academic excellence having first rank merit certificate awardee for his Master's degree programme at the university level. He has organized 5 trainings, 4 workshops and 2 International symposiums etc. He has attended 4 trainings, 2 workshops and 14 seminars/ symposiums/ conferences etc.

Dr. Kumari Rashmi started her early career in the Department of Plant Breeding and Genetics at Bihar Agricultural College, Bihar Agricultural University, Sabour, Bhagalpur in 2012 as an Assistant Professor-cum-Junior Scientist and worked extensively on speciality Corn breeding and heat stress tolerance in Maize in the state of Bihar. She has dedicated more than eight years of her career to agricultural teaching, research, extension, training and administration. She has experience in handling 5 projects including international collaborations.She handled a USAID funded international project on Heat Stress Tolerant Maize based on Genomic Selectionas Co-PI. She also handled various University-funded projects on Speciality Corn breeding and Maintenance Breeding as PI.She has guided 5 M.Sc. students and 1 Ph.D. student as a Major advisor. She has also trained several Master's and Ph.D. students in Maize Breeding. She has published more than 30 research papers, 45 abstracts published in proceedings and souvenir, 13 pamphlets as well as edited and contributed to more than 4 books and 2 book chapters. She has contributed in development of a short-duration variety of maize for the kharif season released from CVRC. She is also a Life Member of the Indian Society for Genetics and Plant Breeding (2012), Society for Scientific Development in Agriculture & Technology, Meerut (U.P.) India (2012) and Society for Agriculture Innovation and Development (SAID),

Ranchi, Jharkhand, India (2017). She was conferred with the Young Scientist Award (2013, 2014 & 2015), Fellowship Award(2014)from the Society for scientific development in Agriculture and Technology, Meerut, UP, India. She was conferred with the Young Scientist Award (2017) in the International Conference-Global Meet on Science and Technology for Ensuring Quality Life (GMST-2017), Kualalampur, Malaysia from Hi-Tech Horticultural Society Prerna Foundation. She was conferred with the Young Teacher Award (2017) and Young Scientist Award (2017) in the International Conference-Agriculture & Applied Sciences for Food Security (AAPS-2017), Kathmandu, Nepal.

Integrated Farming System Models
Development and Implementation

Adyant Kumar
Assistant Professor cum Junior Scientist
Department of Agronomy, Dr. Kalam Agricultural College
(Bihar Agricultural University), Kishanganj, Bihar

Sudhanand Prasad Lal
Assistant Professor
Department of Extension Education, Post Graduate College of Agriculture
(Dr. Rajendra Prasad Central Agricultural University), Pusa, Samastipur, Bihar

Milind B. Katare
Subject Matter Specialist (Fisheries)
Krishi Vigyan Kendra (Central Agricultural University, Imphal)
East Garo Hills, Meghalaya

Reena Kumari
Assistant Professor
Department of Agricultural Engineering, N.M. College of Agriculture
(Navsari Agricultural University), Navsari, Gujarat

Pratibha Kumari
Subject Matter Specialist (Soil Science)
Krishi Vigyan Kendra
(Bihar Agricultural University) Araria, Bihar

D.K. Jayswal
Assistant Professor cum Junior Scientist
Department of Horticulture (Fruit & Fruit Technology)
Bihar Agricultural University, Sabour, Bhagalpur, Bihar

Kumari Rashmi
Assistant Professor cum Junior Scientist
Department of Genetics and Plant Breeding, Dr. Kalam Agricultural College
(Bihar Agricultural University), Kishanganj, Bihar

NIPA® GENX ELECTRONIC RESOURCES & SOLUTIONS P. LTD.
New Delhi-110 034

NIPA® GENX ELECTRONIC RESOURCES & SOLUTIONS P. LTD.

101,103, Vikas Surya Plaza, CU Block
L.S.C.Market, Pitam Pura, New Delhi-110 034
Ph. +91 11 27341616, 27341717, 27341718
E-mail: newindiapublishingagency@gmail.com
www: www.nipabooks.com

For customer assistance, please contact
Phone: + 91-11-27 34 17 17
Fax: + 91-11-27 34 16 16
E-Mail: feedbacks@nipabooks.com

Print ISBN: 978-93-58871-98-2

ebook ISBN: 978-93-58874-32-7

Composed and Designed by NIPA®.

BIHAR AGRICULTURAL UNIVERSITY, SABOUR

BHAGALPUR - 813 210 (BIHAR) Mob.No. : 7070425151

Dr. D. R. Singh
Vice Chancellor

Phone : 0641 - 2452606 (O)
Fax : 0641 - 2452604
Patna : 0612 - 2222267 (O)
Fax : 0612 - 2225364
E-mail ID : vcbausabour@gmail.com

Ref. 354/VC

Date : 01. 12. 2023

Foreword

India has a rich diversity of livestock, poultry, crops and horticulture. Utilising our natural resources efficiently is very important for sustainable development. Integrated farming system (IFS) is recognised as a solution to mix farm enterprises such as crop, livestock, aquaculture, poultry, sericulture and agroforestry to achieve economic and sustained agricultural production through efficient utilisation of resources to fulfil the continuous increase of demand for food production, providing stability to the income and nutritional security particularly for the small and marginal farmers with limited resources as the small and marginal farmers are the core of the Indian rural economy constituting 85% of the total farming community.

This book covers the different IFS models developed, such as wastes generated from one component becoming an input for another system, leading to efficient recycling of farm and animal wastes in the integrated system. Moreover, IFS increases yield and productivity per unit area through the intensification and diversification of crops and allied enterprises. The integration of different production systems provides an opportunity to solve malnutrition problems in our country. Apart from this, IFS helps in controlling insect pests, diseases and weeds through natural cropping system management and there is less use of harmful agrochemicals for farm production. Therefore, it improves soil fertility and physical structure through appropriate crop rotation and the use of cover crops and organic compost.

IFS provides regular stable income through products like eggs, milk, mushroom, vegetables, honey and silkworm cocoons from the linked activities in integrated farming, producing higher net returns to the land and labour resources of the farming family by reducing the production costs of components through input recycling from the by-products of allied enterprises. Thus, this system of farming is very promising for improving overall farm productivity, profitability, generating employment opportunities, conserving natural resources and maintaining the sustainability of the agroecosystem by effectively recycling farm by-products and efficiently utilising available resources.

I wholeheartedly congratulate the team and appreciate the themes of this book. The book will show directions to students, researchers, farmers, and policymakers towards the future prospects of an integrated farming system. I am also sure that it will go a long way in helping students succeed in research projects of UG, PG and Ph.D. and attract bright students to IFS research, education, and extension. My best wishes to the team for bringing out this publication.

Dr. D.R. Singh
Vice Chancellor

Preface

Embodying global agricultural prominence, India's contribution to the GDP stands strong at 18.30% in the year 2022-23. It claims unrivalled leadership in milk, pulses, and spices, while securing the second spot in a plethora of essentials, including rice, wheat, cotton, sugarcane, and more. With India's population projected to reach 1.50 billion by 2035, the demand for sustenance is on a resolute rise, with food grains projected to increase from 315 MT (2022) to 334.90 MT (2025). Consequently, the nation faces a pressing need to amplify its agricultural output to meet these demands. However, a notable decline in average landholdings is observed, diminishing from 1.08 ha in 2015-16 to 0.68 ha in 2020, signalling a reduction from 1.41 hectares in 1995-96 to 0.32 ha in 2030 if current trends persist, as unveiled by the Agriculture Census Report (2015-16). With a gradual declining trend in the size of landholding, it poses a serious challenge to the sustainability and profitability of the farming community because of the execution of an unorganisedand ineffective farming system and unutilised resources. The government hasalso introduced many schemes to upgrade the farming community and increase income but due to the fragmented landholding and conventional practices, farm mechanisation and new technologies are not able to be implemented. This situation in India calls for an integrated effort to address the emerging livelihood issues. Thus, it is imperative to develop strategies and agricultural technologies that enable adequate income and employment generation for small and marginal farmers', who constitute more than 85 per cent of the farming community. Therefore, in today's farming context, a system or approach should be suitable and considered the most powerful tool for the fragmented landholding, cutting down input costs, providingfor the reuse of waste as input for another and enhancing the profitability of small and marginal farmers. The integration of enterprises leads to greater dividends than single enterprise-based farming, especially for small and marginal farmers. From aneconomic aspect, IFS may improve the economy of small and marginal farmers as production and costs decrease by optimising resource utilisation and recycling waste materials.

While possessing numerous advantages, the implementation of Integrated Farming Systems (IFS) at the smallholder level remains limited. Small and marginal farmers encounter obstacles when trying to integrate production systems, agri-business management, and risk mitigation within the IFS framework. Successful IFS operation demands individuals driven to create or expand economic activities, exploring new products, processes, and markets. This hinges on an innate willingness to innovate, embracing calculated risks for future gains and sustaining a business with an entrepreneurial mindset. Entrepreneurial behaviour, encompassing traits like innovativeness, decision-making processes, and risk orientation, plays a pivotal role. Recognizing the significance of these factors is pivotal in cultivating an environment that nurtures entrepreneurial behaviour. Tailoring IFS strategies, competencies, and management to align with agricultural entrepreneurship's essence becomes paramount in this pursuit.

Entrepreneurship within agriculture assumes paramount significance as it entails devising novel strategies to harness opportunities unattainable by others. This involves the adept fusion of processes, practices, and decision-making approaches that imbue organisations with entrepreneurial vigour. Integrated farming is a versatile concept adaptable to varying situations and land types, without adhering to a fixed pattern. It offers lucrative prospects for youth, empowering them to plan effectively and evolve as entrepreneurs while generating rural employment. Moreover, it emerges as an attractive avenue for women entrepreneurs due to its cost-efficiency and potential for substantial income.

With this background, the book has been designed to discuss the aspects and elements of entrepreneurship that must be present in developing integrated farming system models and theirsuitability under different agro-climatic situations. "Integrated Farming System Models: Development and Implementation" likely explores the concept of entrepreneurship within the context of integrated farming systems. An integrated farming system involves the combination of various agricultural activities such as crop cultivation, animal husbandry, aquaculture, agroforestry, lac culture and more, with the aim of optimising resource utilisation, increasing productivity, and achieving sustainability. In the book, one can expect to find discussions on the introduction to integrated farming, entrepreneurship in agriculture, business planning, value chain integration, sustainability and environmental considerations, case studies, challenges and opportunities, policy and support, and future trends.

We trust that this book offers a varied and appropriate array of Integrated Farming System (IFS) models, grounded in research discoveries and policy recommendations. These insights can prove invaluable for policymakers, financial establishments, researchers, and anyone seeking to recognise the obstacles and inventive solutions within integrated farming systems, all aimed at contributing to the Vocal for Local Mission, self-sufficiency and enhancing the livelihood security of Indian farmers.

Editors

Contents

1

Integrated Farming System (IFS) - A Holistic Approach to Empower Marginal Farmers

Suprava Biswal[1] and Toko Yarin[2]

[1]*Department of Vegetable and Spice Crops, Uttar Banga Krishi Viswavidyalaya, Pundibari, Cooch Behar, West Bengal*
[2]*Department of Plantation Crops and Processing, Uttar Banga Krishi Viswavidyalaya, Pundibari, Cooch Behar, West Bengal*

Abstract

The term "Integrated Farming System" (IFS) refers to a set of interdependent, inter-related and frequently interlocking production systems that are based on a small number of crops, animals and related subsidiary ventures. It is designed to maximise the utilisation of nutrients in each system and minimise the negative environmental impact. These are essentially integrated land-based ventures that can grow different types of horticultural crops at different times of the year, along with annual dairy and poultry, using low-cost farming techniques to improve job creation through sustainable production. It not only helps farmers in income generation but also encourages them to focus on using on and off-farm units. In this system farmers can get benefit throughout the year by giving opportunities to other tribal farmers including females. Since farming is a labour-intensive work, in many places the involvement of females plays a very important role in women's empowerment. In present days, many tribal females in Rajasthan and Punjab are getting benefit more from their farming system which helps creating jobs for other indigenous women. Farming various horticultural crops such as vegetables (summer and Rabi types), fruits, plantations, medicinal and aromatic crops by using low-cost farm technology i.e., incorporation of organic farming at minimum input costs can bring tremendous opportunities for income generation. The main objective of this study is to bring more benefits for small marginal farmers as well as open up new avenues for the higher income generation.

***Keywords*:** IFS, Ventures, Sustainable, Tribal, Income generation, Marginal farmers.

Introduction

The total geographical area for farming is 328.7 million hectares whereas, only 139.4 million hectares are a net sown area in India (NHB). India is known as an agricultural powerhouse where small and marginal farmers are at the centre of the Indian rural economy. States like Punjab and Haryana have 80% of the net sown area whereas, Arunachal Pradesh, Mizoram and Manipur have only 10% of the net sown area. With the population increasing every year, innovative farming method should be used to feed the large population, utilising vertical space to make the most limited space and time.

After the innovation of new farming techniques with useful farming tools, the Indian farming system needs a lot of man power for carrying out all the farming operations. Despite working hard, farmers often struggle to sustain their livelihoods due to the high costs of farming inputs such as seeds, manures, pesticides, energy, and feed especially for small and marginal farmers. To satisfy the fundamental necessities of these homestead families including food (cereals, pulses, oilseeds, milk, organic product, honey, meat and so forth), fibre and fuel warrant a consideration about incorporated cultivating framework (integrated farming system).

Integrated Farming System (IFS) refers to a set of interdependent, interrelated and frequently interlocking production systems that include a small number of crops mainly involving the cultivation of various crops, animals and related subsidiary ventures. IFS is just a simple term which is generally used in the agriculture sector to define agro-ecological practices, land productivity as well as environmental sustainability to improve farming system for farmers. In this system, an interdependent set of enterprises is formulated, which means waste from one component can be used as a source for another component of farming which will help in development of low-cost farming technology for farmers.

After the Green Revolution in the years 1965-1966, farmers are mostly focused on single farming enterprises. In a single enterprise, using a lot of inorganic fertilizers, insecticides and pesticides which cause degradation of soil quality and as well as hazardous to the environment. Sometimes the failure of a single enterprise is more due to climatic effects, today's demanding economy, consumer preferences, the spread of harmful diseases etc. But if a marginal farmer focuses on multiple enterprises, that could give better livelihood with employment generation for a small-scale farmer. As a multiple enterprises or multiple farming which demands heavy man wages, women can be a better option for carrying out all intercultural operation that could help in empowering the woman. Food is in greater demand in the modern world due to growing population. Hence, integrated farming system is a tool used for empowering the farmers with improving livelihoods.

Goals of Integrated Farming System

(1) To support continuous and steady income generation crop-based system productivity.

(2) To balance the agro-ecological system by reducing the population of insect-pests, introducing a natural farming system by avoiding the use of inorganic fertilizers, insecticides and pesticides.

(3) To implement technology that judiciously uses of local resources available on the farm and in local areas effectively and efficiently.

(4) To promote long-term prosperity, maintain the environment, and preserve natural resources.

Advantages of Integrated Farming System to Empower the Marginal Farmers

a) IFS makes the system more feasible by combining a complex matrix of soil, water, plants, animals, and the environment with their interactions more profitable than the arable farming system and increases the output of wholesome food.

b) The revenue generated by crops is hardly enough to keep the farm family afloat throughout the year.

c) When the crop is in good condition, a guaranteed regular cash flow is merged with other enterprises, an intelligent mix of enterprises mindful of the environmental circumstances of a locality will yield bigger rewards. While doing so, additionally, it will support efficient recycling of waste.

d) On the same plot of land, horticultural and vegetable crops can provide up to three times as much energy as cereal crops, ensuring the sustainability of the current system's food supply and economic viability (Gill *et al.*, 2009).

e) Due to the need to conserve space, the addition of beekeeping, fishing, mushroom growing, bird rearing, goatry, and livestock also provides additional high energy without compromising the production of food grains. The integration of these businesses promotes realistic production, consumption, and breakdown within an ecosystem. Since biomass and animal waste still contain between 70 and 80 percent of the micronutrients that are present in crops, proper recycling of waste is essential in farming systems (Inman *et al.*, 2005).

Components of Integrated Farming System to Empower the Marginal Farmers

Integrated farming systems can be classified into four major groups as mentioned below –

a) **Crops:** Horticultural crops like different types of vegetables (legume vegetables, bulb types, cucurbits, solanaceous etc.), fruit crops (mango, citrus, guava and temperate fruits), ornamental and foliage plants, plantations, spices, medicinal aromatic plants, and fodder crops.

b) **Livestock and Poultry:** Cows, buffaloes, goats, ducks, hens, pigs, sheep etc.

c) **Fishery:** Poultry unit is set up above the pond where fish can get food and ultimately give an income generation for marginal farmers.

d) **Secondary Farming:** Mixed farming or intercropping systems mushroom cultivation, bee keeping, biogas plants, vermicomposting etc.

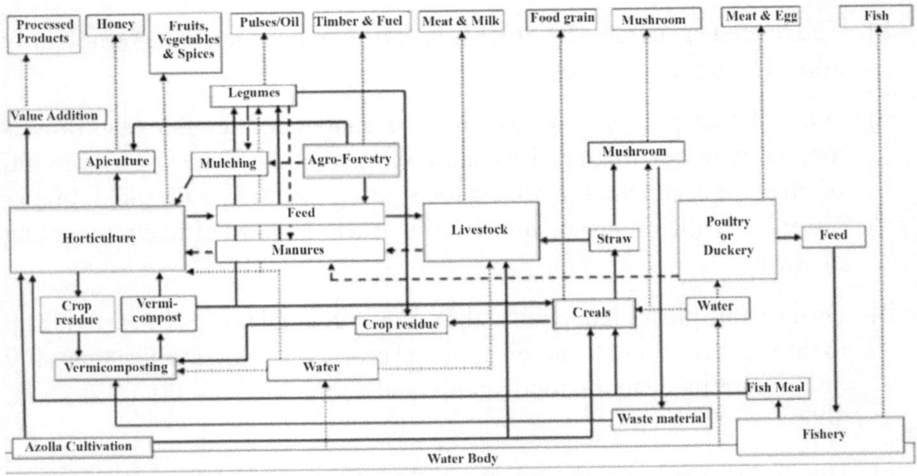

Fig. 1: Different components of Ideal IFS model and their interaction

Energy Transfer Model to Empower Farmers

In the present scenario where people are very much health-conscious and concerned about their food, they mostly prefer organic and fresh. The trend has led to competition among big companies to push themselves to produce more organic vegetables, fruits etc. to the customers, which is in fact opening doors for small businesses to the farmers for income generation. In India most of the farmers apart from crop cultivation, have their own livestock which could be helpful for them in many ways such as producing organic manures vermicompost along with organic fruits, vegetables etc. for income generation. In recent years, the demand of organic manure is also increasing tremendously in markets. There are many ways available from on and off farm units, which can be used for manure production.

Manures

Manures are the organic materials obtained from human, animal and plant residues that are used as sources of plant nutrients to improve crop productivity (Figure 2). Manures with low nutrient content per unit quantity have a longer residual effect apart from improving soil physical properties compared to fertiliser with high nutrient content. Manures are categorised into bulky organic manure and concentrated organic manure.

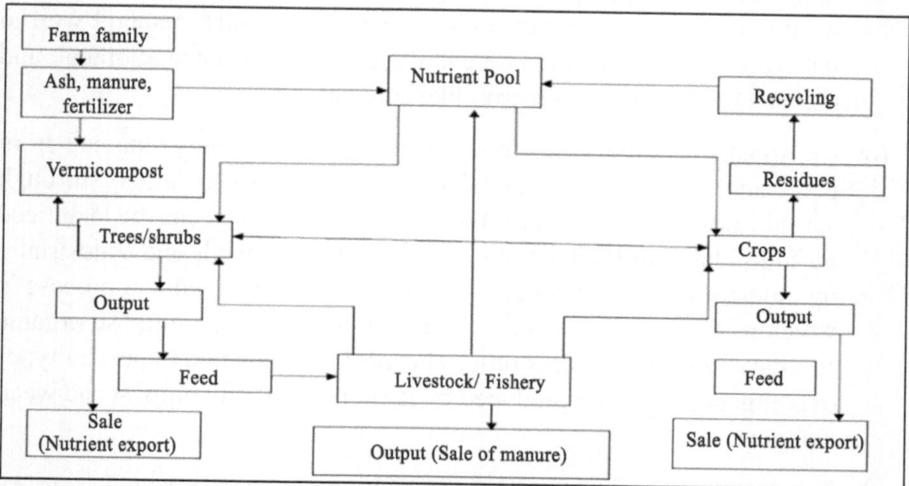

Fig. 2: Flow chart of major sources of manures (*Source*: Kumar *et al.*, 2018)

1) Bulky Organic Manures

The most important and extensively used bulky organic manures include Farmyard manure (FYM), compost and green manure which can be produced by the farmers themselves. All the materials required for the preparation of FYM, compost and green manures can be obtained from the waste of cows, goats, sheep, crop residues as well as weeds around the field. All these compost materials are cheap and available to farmers which they can turn into organic manure and sell in the market. The advantages of bulky organic manure are –

- They supply plant nutrients including micronutrients.
- They improve soil physical properties like structure and water holding capacity.
- They increase the availability of nutrients.
- Carbon dioxide released during decomposition acts as a CO_2 fertiliser.
- Plant parasitic nematodes and fungi are controlled to some extent by altering the balance of microorganisms in the soil.

a) **Farmyard manure** – It refers to the composition mixture of dung and urine of farm animals along with the litter, bedding material like hay, straw used for cattle or fodder fed to the cattle. A well-rotted FYM contains an average of 0.5% N, 0.2% P_205 and 0.5% K_20 (Anonymous). Some of the horticulture crops such as vegetable crops viz. tomato, potato, sweet potato, carrot, radish, onion etc., which respond well to farmyard manure as well as orchard crops such as banana, orange and mango and also plantation crop like coconut.

b) **Compost** – Compost is composed of organic materials obtained from plant and animal matter that has been decomposed largely through aerobic decomposition. The process of composting can be practiced by individuals in their homes, farmers on their land, and industrially by industries as well as cities. Composting is a bio-chemical process in which microorganisms both aerobic and anaerobic decompose organic residue and lower the C: N ratio. The most common and important types of compost that can be produced by farmers are vermicompost and weed compost.

c) **Vermicompost** – Since vermicompost is a familiar term which is known to many farmers and common people to improve their farming system. The term is known as the composting of weeds or any plant residue with cow dung by using earth worm to create a heterogeneous mixture. This is the most adopted technology to produce organic manure. Some of the important worms used for this technique are Red Wigglers (*Eisenia fetida* or *Eisenia andrei*), though European night crawlers (*Eisenia hortensis*) can also be used. The unit can be prepared according to capacity. It can be prepared from using low-cost available materials like bamboo, thatches or paddy straw etc. The following are the steps to carry out this method –

- First construct a pit with bamboo of 3.5 m × 2.1 m × 1.5 m size with a thatched roof on top.
- After that chop the weeds and place alternate layers of dung and chopped weeds in the pit. Then let the mix to decompose for one month.
- Now transfer the decomposed material into another pit of 3 m × 1 m × 0.5 m which could accommodate about 500 earthworms and sprinkle water regularly to avoid desiccation of the earthworm.
- A 4-inch-wide drainage should be kept all around the pit with water filled to protect the earthworm from any predators like ants and insects.
- In warmer condition it takes about 3 months to produce first batch of vermicompost.
- The upper layer should be gently removed without harming the worms and allow them to move downwards.
- After collection from the sieve, prepare it for use in the field or sell in the market.

Currently, the cost of 1 kg of varies between ₹6-20 depending on the area as well as producer; whereas, the cost of vermicompost in renowned companies is more than ₹200. Therefore, the small-scale farmers can contact the big companies with the help of KVK's or the Agriculture or Horticulture department.

d) **Weed composting** – Apart from vermicompost, farmers can go for another method of compost by using weeds compost which is cheaper and less time consuming even though the compost from both techniques serves the same purpose. Since mulching weeds is a common practice among rural farmers, weed composting can be developed using low-cost technology to produce high-quality manure that can be used as organic manure by farmers themselves to produce organic product or to sell in the market to earn extra income. In this method the materials required to set up a composting unit are bamboos, thatches/CGI sheets and perforated pipe/bamboo pipes. The composting unit can be constructed by making a three chambered (1m × 1m × 1m) structure with bamboos by placing perforated pipes/bamboo pipe vertically in middle of the first and second chamber for the aeration with thatches/CGI sheet. After completing the construction of compost unit, the following steps can be carried out for weed decomposition.

- Firstly, collect fresh cow dung and fresh weeds/ field or vegetable waste in equal amount (1:1). The weeds should not be in the flowering or fruiting stage. Also avoid any kind of weed that has a sweet aroma or is sour in taste (Figure 3).
- Then chop the weeds into small pieces and mix them thoroughly with fresh cow dung with foot and add a little water for the consistency of the paste.
- After mixing, transfer the paste into the first chamber and keep it there for 9 days to decompose.
- After 9 days, transfer the paste from first the chamber to the second chamber and keep it there for another 7 days.
- Next, transfer the paste into the third chamber and keep it there for another 5 days for complete decomposition. After that, sieve it and it is ready for use. It takes a total of 21 days to complete this procedure. This method is the simplest and cheapest technology for farmers to produce organic manure on their own. According to farmers in the hilly areas of Arunachal Pradesh, weed compost has shown excellent results for the growth of vegetables as well as other crops.

Fig. 3: Showing low cost weed composting unit and demonstration of mixing fresh weed and cow dung

2) Concentrated Organic Manures

Concentrated organic manures are naturally organic in nature and contain higher percentages of major plant nutrients like N, P_2O_5 and K_2O compared to bulky organic manures. These concentrated manures are produced by using raw materials with either an animal or plant source such as oil cakes, blood meal, fish dung, meat meal, and wool waste.

Successful Model of Integrated Farming System

Kerutagi *et al.* (2019) conducted a study in the year 2016-17 in Northern Karnataka using 4 types of horticulture-based cropping systems. In this system, a total of 80 farmers were involved in carrying out the integrated farming system and non-integrated farming system (farm size-2.80 ha.) to determine the effectiveness of the experiment (Figure 4 and 5).

IFS-1 (Farm size – 3.50 ha): It involved a farming system of mango, banana, coconut, sapota, vegetables, sugarcane, pulses, maize, flowers, livestock like goat, dairy, poultry and vermicompost.

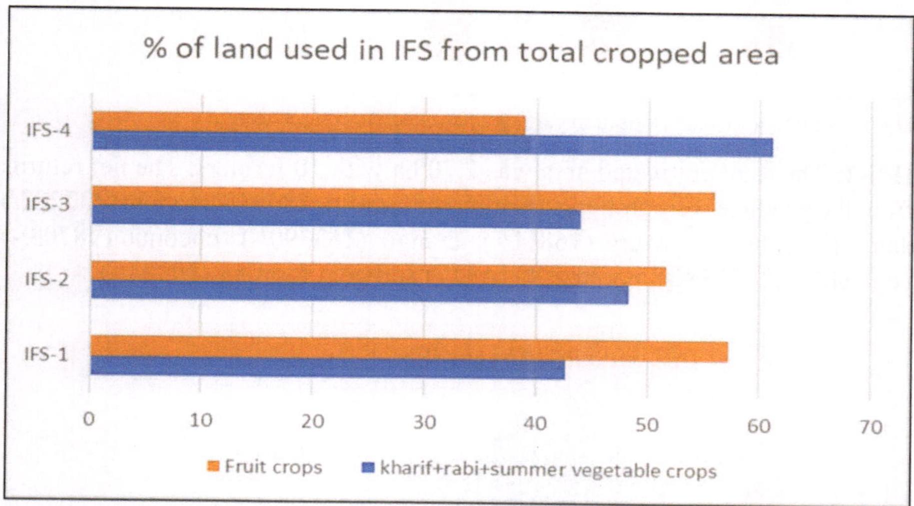

Fig. 4: Cropping % of land area used IFS-1 (Farm size – 3.50 ha)

IFS-2 (Farm size – 2.70 ha): It included all crops present in IFS-1 except pulses.

IFS-3 (Farm size – 2.4 ha): It involved the farming system of mango, banana, coconut, grapes, sapota, onion, chilli, maize, livestock components like goatry dairy, poultry and vermicompost.

IFS-4 (Farm size-3.05 ha): It involved the farming system of mango pomegranate, sapota, grapes, drumstick, chilli, onion, maize, goatry, poultry dairy and vermicompost.

Economics of Different Farming Systems (Kerutagi *et al.*, 2019)

IFS-1: The cultivated area of IFS-1 was 3.50 ha. The total number of farmers were 10. The net return of the produce was as follows: vermicompost unit (₹17466/-), mango (₹67754/-), dairy unit (₹15679/-), sugarcane (₹22640/) vegetables (₹10927/-), poultry (₹6396/-), flowers (₹3581/-) and sapota (₹20160/-) etc.

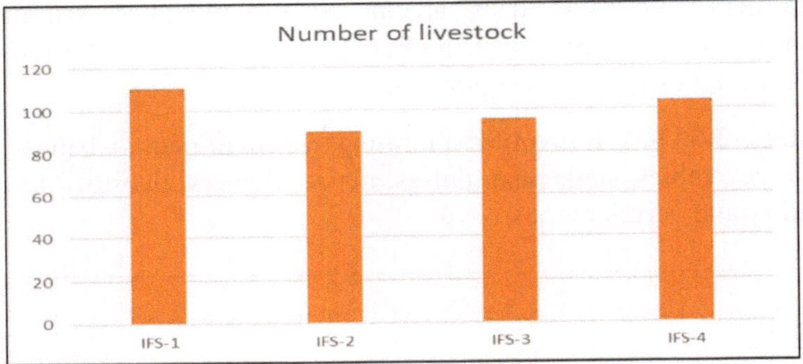

Fig. 5: Number of livestock used in various Integrated Farming Systems

IFS-2: The total cultivated area was 2.70 ha with 10 farmers. The net returns from the produce were from the vermicompost unit (₹8094/-), sapota (₹37282/) dairy (₹32111/-), flowers (₹6811/-), mango (₹63390/-), coconut (₹8703/-) vegetables (₹15788/-), poultry (₹11430/-) and goat farming (₹9384/-).

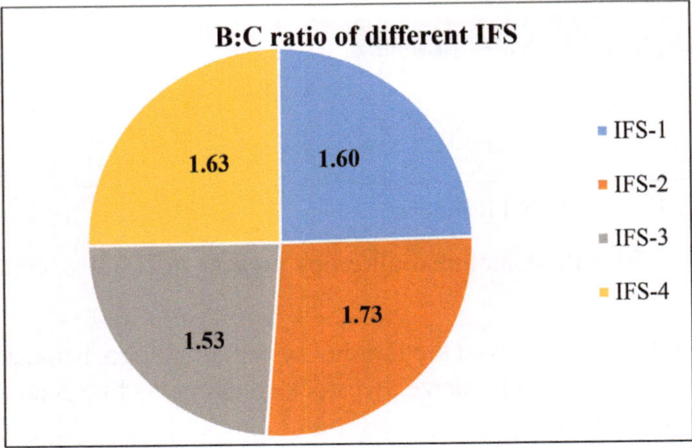

Fig. 6: B:C ratio of different IFS

IFS-3: The total cultivated area was 2.40 ha with 10 farmers. The net return of the produce was from vermicompost unit (₹9372/-), mango (₹61521/-), goat farming (₹15931/-), dairy (₹21505/-), grapes (₹113918/-), chilli (₹12896/-), banana (₹12,279/-), onion (₹8566/-), sapota (₹30752/-) and coconut (₹5149/-) etc.

IFS-4: The total cultivated area was 3.05 ha with 10 number of farmers. The net return of the produce obtained from vermicompost unit (₹10650/-), mango (₹92492/-), pomegranate (₹65818/-), dairy (₹18555/-), poultry (₹9214/-), sapota (₹53158/-), chilli (₹17148/-), goat farming (₹7032/-), onion (₹13927/-) and grapes (₹53158/-).

Conclusion

As the nation's population is increasing day by day, it is directly affecting the environmental conditions. To protect the sustainability of the environment, the adoption of an integrated farming system plays a very important role as compared to non-integrated farming systems. Since this system is interdependent with ecology, it improves the living system of the earth. As per the data interpretation, it has been found that IFS is showing a positive effect on farmers as well as industries/ cities that have adopted this system with great economic returns with the increase in the curve of Benefit Cost (BC) gain per unit expenditure. The economic section of the organisation has also immensely improved due to the integration of more than three enterprises. Integrated farming systems have benefited many small-scale farmers apart from the production of vegetables, fruits etc. by producing organic manures which has a great impact on their earning system. It also benefitted employment generation in rural areas and women empowerment by giving them opportunities to take part in entrepreneurship and boosting their self-confidence. Therefore, one should encourage the adoption of this system to bring about a change in their living style.

References

Gill, M.S., Singh, J.P. and Gangwar, K.S. (2009). Integrated farming System and agriculture sustainability. Indian Journal of Agronomy. 54(2): 128-139.

Inman, D., Khosla, R., Westfall, D.G. and Reich, R. (2005). Nitrogen uptake across site- specific management zones in irrigated corn production systems. Agronomy Journal. 97: 169-176.

Kerutagi, M.G., Talavar, M. and Pavitra A.S. (2019). Impact of horticulture based integrated farming system on farmer's income and welfare in Northern Karnataka. Journal of Pharmacognosy and Phytochemistry. 8(3): 1010-1019.

Kumar, S., Bhatt, B.P., Dey, A., Shivani, Kumar, U., Idris, Md., Mishra, J.S. and Kumar, S. (2018). Integrated farming system in India: current status, scope and future prospects in changing agricultural scenario. Indian Journal of Agricultural Sciences. 88 (11): 1661-1675.

2

Horticulture
A Viable Alternative for Sustaining Integrated Farming System

Kajal Sengupta and Prasun Chatterjee

Division of Agronomy, Faculty of Integrated Rural Development and Management (IRDM), School of Agriculture and Rural Development Ramakrishna Mission Vivekananda Educational and Research Institute [Deemed to-be-University], Narendrapur Campus, Kolkata, West Bengal

Abstract

Horticulture can be regarded as a major agricultural practice in the entire nation which mainly aims at providing an opportunity to increase farmer's income and raise their standard of living by increasing the land's area and productivity through proper utilisation of technology. Farm mechanisation technologies and the development of new HYV crop varieties/hybrids became the area of priority for researchers in the past few decades, which led to the over-exploitation of natural resources thus, completely overlooking the environmental aspect. Continuous land degradation with the removal of topsoil due to erosion and nutrient mining due to the growth of major cereal crops every year leads to low productivity and prevents farmers from getting a good price for farm produce. Here, IFS plays a major role whose principle mainly deals with the integration of different enterprises along with crop production where the output of one component such as waste materials and products turns out as an input of another component with the reduction in the production cost. Horticulture based IFS model emerges to be a promising one as the cultivation of horticultural crops can be a major source of income generation and food security. These horticulture-based IFS models offer great scope for entrepreneurial activities, especially in irrigated regions and much of it needs to be explored under varied agro-climatic zones.

Keywords: Integrated farming system (IFS), Components, Enterprises, Doubling income.

Introduction

We live in India, a country which is in its developing phase and would be developed soon and also where most of the population is concentrated in the rural areas. So, eventually, rural development emerges to be of much importance for the overall economic development of the nation. But it's a big disgrace for us that the population in the majority lies below the line of poverty. The development of households in rural areas is of much significantly related to entrepreneurship. Entrepreneurship can be regarded as a tool to fasten up the process of rural development. An individual who effectively and competently combines the four factors of production is regarded as an entrepreneur. Those factors of production may be categorised as labour or human input into production, land or natural resources, capital or equipment that is used in products such as machinery, and lastly creativity, knowledge, and intelligence. The major problem that lies is not in conducting research work but rather to disseminate the research findings to the farmers probably due to poor extension services. As a solution to this, Agriculture Entrepreneurship Development Programmes (EDPs) have been created which aim at disseminating the recent developments from the laboratory to the land.

Horticulture can be defined as a science that deals with the cultivation of gardens majorly of vegetables, fruits, flowers, and ornamental plants. Horticulture mainly aims at providing an opportunity of increasing the farmer's income as well as raise their living standard with increment in the land's area and productivity by proper utilisation of technology. There has been a diverse opportunity for growing many horticultural crops because of the varied agro-climate in India. It is categorised as the second-largest producer of fruits and vegetables (Sambrani, 2016). The nodal agency in India which performs the task of encouraging horticultural entrepreneurship in the country is the Department of Agriculture and Co-operation under the Ministry of Agriculture. Formulation of various schemes like the "National Horticulture Mission" and "Technology Mission for Integrated Development of Horticulture" has performed the role of deriving numerous benefits for diverse horticultural crops in the various regions of North India. National Horticulture Board (NHB) was set up to promote horticulture for proper coordination and production of plants for vegetables and fruit processing.

Farm mechanisation technologies and the development of new HYV crop varieties/hybrids became the area of priority for researchers, starting from the green revolution till last few decades. This led to the over-exploitation of natural resources thus, completely overlooking the environmental aspect with decrement in much of productivity and also in other hand profitability.

Continuous land degradation with the removal of topsoil due to erosion as well as nutrient mining due to the growth of major cereal crops every year with the influence of climatic change lead to low productivity as well as prohibits farmers from getting a good price for farm produce. Here, IFS plays a major role whose principle mainly aims at a combination of varied enterprises of the farm like vegetables, field crops, dairy, poultry, and goatery for a sustainable and profitable farming system. Integrated farming system practices try to supplement the production needs without hampering the social ecological, and economic balance of the system. Amongst various IFS models Horticulture based IFS model emerges to be a promising one as the cultivation of horticultural crops can be of the major sources of income generation and food security. IFS majorly deals with the integration of different enterprises along with crop production where the output of one component such as waste materials and products turns out as an input of another component with a reduction in the production cost. However, horticultural production's large percentage is mainly concentrated in regions with abundant water availability leaving a large scope of its expansion in the hilly regions and also in the rainfed regions which comprise more than 60% of India's production area (Sambrani, 2016). It can be done by de-siltation of tanks, conduction of various watershed development programmes and also increasing moisture content in the soil resulting in improvement in productivity and production.

Case Studies Related to Entrepreneurial Activities in Horticulture

1. The case study depicts the story of a farmer named ***Mr. Manjunath Neginal*** of Karnataka, who independently started his nursery with lots of hardship but his hard work and dedication carved for him a recognisable identity in society. Initially, he started to work despite having his land of 3.5 acres. After a few days, he quit his job and planned to start vegetable farming but the main constraint was to get the saplings from far-off nurseries. At first, he was forced to create its shed nets on small scales and grow the saplings which didn't give proper yields. So, he decided to start his nursery and after two years of learning was profound enough to start his nursery in an area of 4 gunta/ guntha (1 gunta = 0.025 acre) and then expand it to 7.5 guntas on demand from the local farmers. So per demand, every batch approximately consisting of 3 lakh saplings, leading to around 24 lakh saplings a year were produced and sold with an approximate income generation of ₹15000/- a month. During the initial start-up, he approached the horticulture department for assistance where he was provided a 50% subsidy for his establishment. At present, he likely saves around ₹ 2.6 lakhs per annum after all his expenses which proved evidence of his consistency in the delivery of performance.

2. Experiments were conducted at the Project Directorate for Farming Systems Research in Modipuram, Uttar Pradesh, India, during 2011-12. The purpose of these experiments was to develop a model based on horticultural crops to improve profitability, productivity, and nutritional security in the western plain zone of Uttar Pradesh. The evaluation included three modules: a vegetable crop-based module (CS 2, 0.22 ha), a field crop-based module (CS 3, 0.4 ha), and a fruit crop-based module (CS 1, 0.3 ha) (Figure 1).

- Under the fruit-based cropping system (CS-1), the main crops cultivated were bananas, guava, and mango. Inter-cropping was done with onion, radish, carrot, and cucumber, and turmeric was intercropped with banana. In guava, intercropping was done with brinjal, vegetable pea, and okra. The vegetable-based system (CS-2) involved the cultivation of brinjal-potato, turmeric, and bottle gourd-cauliflower-tomato. The crop-based system (CS-3) evaluated sugarcane-ratoon and rice-wheat.
- After evaluation, it was observed that the crop-based system yielded lower benefits and returns compared to the vegetable-based system, which had higher returns and benefits. In the second year, the vegetable-based system (CS-2) generated a net return of ₹151,640/ha, followed by the crop-based system (CS-3) with a return of ₹68,765/ha, and the fruit-based cropping system (CS-1) with a return of ₹44,660/ha. The higher returns in the vegetable-based system were likely due to the cultivation of cash crops, while the lower returns in the fruit-based system could be attributed to the absence of returns from fruit plants in the initial years.

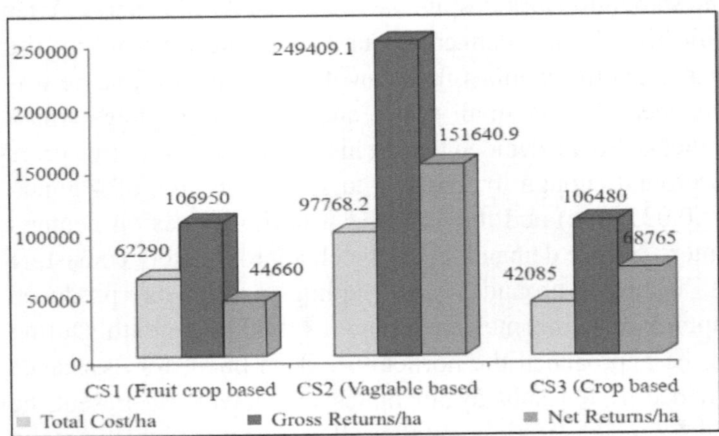

Fig. 1: Comparative study on the economics of different cropping system [Source: Kashyap *et al.*, 2015]

3. Based on a horticulture-based integrated farming system, an experiment was carried out in Mudhol taluka of Bagalkot district and Gokak taluka of Belagavi district, Karnataka. The entire experiment was divided into four farming system models comprising a combination of various components including field crops, horticulture, dairy, and vermicomposting, favouring residue recycling and hence reducing the cost of cultivation.

- **FS-I:** It comprises banana, coconut, mango, vegetables, sugarcane, pulses, sapota (*Manilkara zapota*), maize, flowers, goat farming, poultry, dairy and vermicomposting. Under IFS-I (Table I), out of the combination of various enterprises, the maximum net returns (₹67754/-) were observed against mango, followed by sugarcane (₹22640/-).

Table 1: Evaluation of economics for different enterprises in IFS-I [Source: Kerutagi *et al.*, 2019]

Sl. No.	Economics	Amount (₹)
1)	Total cost of cultivation	332863
2)	Gross return	531649
3)	Net return	198756
4)	Benefit: Cost ratio	1.60

FS-II: It comprises a system of mango, banana, vegetables, coconut, flowers, sugarcane, sapota (*Manilkara zapota*), maize, dairy, vermicomposting, poultry, and goat farming. Under IFS-II (Table 2), out of a combination of various enterprises, the maximum net returns (₹63390/-) were observed against mango, followed by sapota (₹37282/-).

Table 2: Evaluation of economics for different enterprises in IFS-II [Source: Kerutagi *et al.*, 2019]

Sl. No.	Economics	Amount (₹)
1)	Total cost of cultivation	342410
2)	Gross return	593561
3)	Net return	251151
4)	Benefit: Cost ratio	1.73

- **FS-III:** It included a system comprising of onion, banana, sapota (*Manilkara zapota*), coconut, mango, chilli, grapes, maize, dairy, poultry, goat farming, and vermicomposting. Under IFS-III (Table 3), out of a combination of various enterprises, the maximum net returns (₹113918/-) were observed against grapes, followed by mango (₹61521/-).

Table 3. Evaluation of economics for different enterprises in IFS-III [Source: Kerutagi *et al.*, 2019]

Sl. No.	Economics	Amount (₹)
1)	Total cost of cultivation	564263
2)	Gross return	860590
3)	Net return	296327
4)	Benefit: Cost ratio	1.53

- **FS-IV:** It comprises a system of drumstick, mango, sapota (*Manilkara zapota*), pomegranate, chilli, grapes, onion, maize, poultry, dairy, vermicomposting, and goat farming. Under IFS-IV (Table 4), out of a combination of various enterprises, the maximum net returns (₹92492/-) were observed against mango, followed by grapes (₹88808/-).

Table 4: Evaluation of economics for different enterprises in IFS- IV [Source: Kerutagi *et al.*, 2019]

Sl. No.	Economics	Amount (₹)
1)	Total cost of cultivation	634025
2)	Gross return	1031499
3)	Net return	397474
4)	Benefit: Cost ratio	1.63

4. A small and resource-poor farmer named ***Shri Saṭta Ram*** lives in the Barmer district of Rajasthan. This area is often hit by drought conditions, resulting in the loss of the entire harvest. So, he thought of developing a fruit-based farming system. At a height of 5 m, he built a small water tank in the middle of the farm for a gravity-based drip system along with a water harvesting pond with a capacity of 10 litres under the subsidy and guidance from the state government. Towards the windward side, he planted five rows of lasora (*Cordia dichotoma*) which were supplied by the institute on the boundary. Gola, a ber variety was planted at a spacing of 6m × 6m with the interspaces planted with mung bean and moth bean. High-value crops like sour lime and pomegranate were planted near the pond. Different fruit species of 800 plants are established. Returns of ₹0.15-0.25 lakh from lasora fruits and about ₹1.0-1.5 lakhs from ber from the third year onwards were achieved.

5. Horticulture-Based Systems of Arid Zone

a) **Agri-Horti System:** Green gram, moth bean, pearl millet, and cluster bean are grown between trees of ber. The alleys of ber are most suited for growing green gram without hampering the yield of both ber and green gram with proper adjustment in spacing. The vegetative vigour of ber plants gets enhanced due to the application of inputs for the ground storey crops which wouldn't be facilitated for sole plantation. This agri-horti system (Table 5) also proves to be more sustainable in adverse weather conditions.

Fig. 2: Ber based Agri-Horti Systems [Source: Singh *et al.*, 2017]

b) **Horti-Pasture System:** Ber can be used for multiple purposes, starting from its leaves being used as animal fodder for sheep, cattle, goats, etc. to its fruit being used as an economical resource. The air-dried leaves are highly palatable. Trees such as Jharber (*Ziziphus nummularia*) and other *Zizyphus* species serve as important feed sources especially in the village areas. A horti-pasture system that is most commonly followed at CAZRI is the combination of Dhaman grass (*Cenchrus ciliaris*) and *Zizyphus mauritiana* providing top feeding material for small ruminants and grass for cattle along with fruits of high TSS content. This system provides approximately 15-25 kg dry leaves of ber along with dry grass fodder of nearly 1.2 to 1.5 tonnes.

Fig. 3: Ber based Horti-Pasture Systems [Source: Singh *et al.*, 2017]

Table 5: Economics of horti- pasture and ber-based agri-horti systems [Source: Singh *et al.*, 2017]

	Spacing	Gross Return (₹)	Net Return (₹)	B:C ratio
Agri-horti system	10 m × 5 m	125600	73100	2.39
Horti-pasture system	6 m × 6 m	70400	42700	2.54

6. Economics along with Production Practices of Horticultural Crops under IFS

While some units of livestock play a major role in the income, horticultural crops had also benefitted the farmers by increasing their livelihood status. The crop which is chosen by the farmers depends on the area to be cultivated whether it is rainfed or irrigated and horticultural crops be it seasonal vegetables or any cash crop or spices tend to supply the farmers with ample income source if properly cultivated with the resources available in the region (Table 6).

a) **Economics of seasonal vegetable crop:** Carrots, cabbage, cauliflower, peas, spinach, beans, tomatoes, and coriander leaves along with other leafy vegetables were cultivated throughout the year on a small piece of land. As per the study, a net income (₹7570) with a B:C ratio (2.09) was obtained. On the other hand, vermicompost making was facilitated with some parts of the vegetable wastes that were cleaned before marketing and some were used as fodder for livestock.

b) **Economics of large cardamom crop:** Large cardamom can be regarded as a high-value perennial spice crop, which is planted in the month of June-July in pits that are spaced at 1.5m and with a size of 30 cm × 30 cm × 30 cm. Seedlings or suckers are planted in the middle of the pit. Mulching is done with dry leaves to control weeds and retain moisture. A net income of ₹63122 with a B: C ratio of 1.70 can be generated with a yield of approximately 5.70 quintals per acre through the cultivation of large cardamom under IFS.

c) **Economics of ginger crop:** The planting materials chosen for sowing should be disease-free, have one or two buds, and weigh 40 to 50 g. The sowing should be carried out in the month of February and harvested in November. After the leaves turn completely dry, the rhizomes used as planting material are harvested. A gross income of ₹53480 and net income (₹22880) was obtained with a B:C ratio (1.75) by following IFS. Vermicomposting can be performed with the leftover harvested plants.

d) **Economics of turmeric crop:** In the month of March, the sowing of turmeric is performed with the crop being harvested in December. The gross income obtained was approx. ₹17842 per acre. A net income of

₹4002 with a B:C ratio of 1.29 can be generated upon following IFS. The leftover plant material is used as an input for vermicomposting. Vermicompost can be prepared with the left over plant material acting as a source of input.

Table 6: Economics of Horticultural Systems under IFS [Source: Roy and Kharga, 2022]

Crop	Area (acre)	Yield (q/ha)	Total cost of cultivation (₹)	Gross return (₹)	Net return (₹)	B:C ratio
Vegetables	0.5	4.90	6939	14509	7570	1.09
Ginger	1.0	17.80	30600	53480	22880	0.75
Large Cardamom	1.0	5.70	90278	153400	63122	0.70
Turmeric	0.5	8.10	13840	17842	4002	0.29

7. Horti-Animal Husbandry-Apiary IFS Model

Mr. Vinod Gopal Barve has been practicing an integrated farming system model called Horti-Animal Husbandry-Apiary under the technical guidance of KVK, North Goa for the past seven years. The integrated farming system model covered a total area of 6 hectares and included 2000 Banana plants, 3000 Black pepper plants, 6000 Arecanut plants, 2 Cinnamon trees, and 5 Nutmeg trees. Additionally, Mr. Barve maintained 4 indigenous cows whose excreta was utilized for the production of organic manure. Furthermore, there were 8 honey bee boxes in the model. The annual production from this model comprised 12.9 tonnes of arecanut, 135 tonnes of banana, 300000 coconut nuts, 4.7 tonnes of black pepper, 15 kg of honey and 3 tonnes of organic manure. The net income generated from this integrated farming system model amounted to ₹40.77 lakhs per annum or ₹6.80 lakhs per hectare (Source: ICAR-Annual Report, 2020).

Fig. 4: Intercropping of arecanut, banana, and legumes, with vegetable crops [*Source*: (Techno Serve, 2018)]

8. During the 2012-13 period at the ICAR Research Complex for NEH Region in Jharnapani, Nagaland, an integrated farming system model was established. This model included horticultural crops (mango, banana, Assam lemon, and seasonal vegetables), field crops (mungbean-maize-toria), a fishery unit, and a piggery unit. Among the field crops, maize had the highest returns during the *Kharif* season, while the yields of mungbean and toria were below expectations due to low survival rates and inadequate rainfall. The major area was dedicated to growing vegetables like bottle gourd, pumpkin, and ridge gourd, which generated a net profit of ₹5890/- within the horticulture components. The fruit crops, particularly the two-year-old Assam lemon, yielded a net return of ₹2359/-. The piggery unit contributed the highest returns among all the components, resulting in an overall net income of ₹24394/-. Only 17 kg of fish were harvested from the fishery component, while the remaining fish were left to grow in the next season. The model provided a benefit-to-cost ratio of 2.35 which was advantageous.

Table 7: Economics of Cropping Pattern under IFS [Source: (Rani, 2015)]

Area	**Crops grown**	**Total Cost of production (₹)**	**Gross return (₹)**	**Net return (₹)**
2 acres	Rice	22000	96000	74000
3 acres	Vegetables, Sugarcane	59000	246000	187000
1 acre	Marigold	50000	140000	90000
1 acre	Sweet corn	5800	38000	32200

A farmer named **Bikei Bhoi** living in the outskirts of Bhubaneswar city in Odisha with his parents has 7 acres of land holdings out of which 2.00 acres are lowland and the rest of the land are medium and upland. He has now undertaken IFS for the expansion of the area under cultivation with cash crops as well as floriculture (Table-VII). The cropping pattern consists of –

- In 2 acres of land, High Yielding/ Hybrid/ Aromatic rice in both *Kharif* and *Rabi* is cultivated,
- In 1 acre, marigold production and raising nursery for marigold is done throughout the year,
- In 1 acre of land, sweet corn is grown sequentially in a year,
- In 1 acre of land, cultivation of sugarcane is done,
- In 2 acres of land, different hybrid vegetables like Okra, Cucumber, Brinjal, Ridge Gourd, Beans, *etc.* are grown throughout the year.

9. An extremely successful farming system model centred on coconuts was implemented at the Integrated Farming System Research Station (IFSRS) in Thiruvananthapuram, Kerala, which is part of the Kerala Agricultural University. This IFS model was initiated during the 2011-12 period. The various components of the coconut-based IFS model included coconut trees on bunds and the surrounding area (1480 m^2), teak trees along the borders (400 m^2), as well as multi-tier cropping with papaya, clove, nutmeg, cocoa rose apple, mango, and spices such as ginger and turmeric (620 m^2). The data collected on yield and economic aspects of the model revealed that, on average, 13,227 coconuts were obtained from an area of 0.2 hectares. The total revenue generated from the model was ₹219982/-resulting in an average net income of ₹58454/-. The benefit-to-cost ratio of the coconut-based integrated farming system model stood at 1.37.

10. The productivity of vegetable units (0.07 ha) in the IFS model, including *Kharif* Okra *cv.* Umang (71.86 q/ha), turmeric *cv.* Palam Lalima (274.29 q/ha), and Rabi vegetables such as broccoli *cv.* Palam Samridhi (34.0 q/ha) cauliflower *cv.* Super Star-71 (86.0 q/ha), and peas *cv.* Azad pea-1 (98.95 q/ha), was found to be satisfactory in terms of overall gross (₹14552/unit) and net returns (₹10730/unit), with a B:C ratio of 3.81. This was achieved by adopting improved technologies (Anonymous, 2008) and utilising irrigation water from rainwater harvesting structures of the IFS model on a need basis rather than relying solely on assured irrigation. By following IFS principles and implementing technological interventions based on land use planning farmers can enhance their farm gains and improve their livelihoods through diversification into vegetable cultivation (Bahera and Mahapatra, 1998; Channabasavanna *et al.* 2009).

11. Puttamallappa, a farmer residing in Mallenahalli village in Tarikere taluka of Chikkamagaluru district (Karnataka), cultivated an orchard consisting of coconut, arecanut, and groundnut. However, the income derived from this cropping system was exceedingly meagre and not economically viable. Subsequently, he received guidance to adopt new high-yielding varieties (HYV) of groundnut such as GPBD-4 and GPBD-5, along with the utilisation of bio-fertilisers, in order to enhance the yields. Following these technical interventions, his income experienced a notable increase of 56.40%, amounting to ₹ 670700 (Table 8).

Table 8. Economic Details of the farmer Mr. Puttamallappa [Source: RKVY Report, June 2019]

1.	Existing IFS model	Horticulture, poultry and dairy		
2.	Bench mark income	Details of other components		
		Component	Yield	Annual Income (₹)
		Arecanut	52 q	17,68,000
		Banana	12 q	26,400
		Coconut	6300 number	75,600
		TOTAL		1870000/-
3.	New components introduced and additional income generated (estimated)	Improvements to irrigation system	System modified	Long term benefits
		Composting	Applied to fields	Long term benefits
		Kitchen garden	To meet family needs	5200
		Multi-storied cropping in arecanut	Established	Long term benefit
		Groundnut with new variety (GPBD-4 or GPBD-5	2 q	5,200
		Total additional income		670700/-
4.	Total annual income after technical interventions			1880400/- (+ 56.4%)

12. In order to examine the impact of organic recycling on the yield of fodder grass and coconuts, experiments were conducted through plantation-based horticulture systems in coastal and island ecosystems of India. The whole experimental plot was divided into four treatments. These included monocropping of coconut with recommended inorganic fertilizers, a coconut-based integrated farming system (CBIFS) consisting of various components along with 50% application of recycled organic biomass, a CBIFS with 100% application of recycled biomass, and a CBIFS with 100% application of recommended inorganic nutrients. Hybrid Bajra Napier Co-3 and Co-4 were assessed as intercrops with different management practices. The combined use of organic and inorganic fertilizers significantly increased the green foliage yield of Hybrid Bajra Napier Co-3 (117 t/ha/year) and Co-4 (115 t/ha/year). The CBIFS demonstrated higher coconut yields compared to monocropping particularly when integrated nutrient management practices were applied. A one-hectare coconut-based IFS produced substantial quantities of coconuts, cow milk, broiler chicken, quail bird eggs, bananas, and peppers. Economic analysis indicated that coconut-based integrated farming is an environmentally sustainable system that yields higher income compared to monocropping. The costs and net returns varied among the CBIFS treatments, with the highest returns observed when 50% organic and 50% inorganic inputs were combined. The majority of revenue derived from coconut, dairy, and poultry components. The coconut-based IFS with organic recycling positively influenced the physical, chemical, and biological properties of the soil, leading to significant improvements in soil health compared to monocropping with inorganic nutrient management.

13. The successful growth and productivity of cinnamon, nutmeg, black pepper, kokum, and clove as mixed crops alongside coconut trees under the aberrant coastal climate were documented by Parulekar *et al.* (2020). Tree spices, such as nutmeg, cinnamon, clove, kokum, and allspice, are highly profitable crops among the various spices. Similar finding were also recorded by Patil *et al.* (1991) and Nagwekar *et al.* (2014). These crops have the ability to withstand the changing climate conditions in the region. They are well-suited as intercrops in coconut plantations. The implementation of mixed cropping with these tree spices resulted in a significant increase in coconut palm yield (Table 9), reaching up to 96% after 26 years compared to previous years' average yield (Nagwekar *et al.*, 2014). Among the intercrops, nutmeg proved to be the most profitable, generating a net return of ₹93578 per hectare when grown alongside coconut, while coconut as a mono-crop had a net profit of ₹25987 per hectare (Nagwekar *et al.*, 2014). The practice of intercropping with nutmeg, kokum, black pepper, cinnamon, and clove in coconut plantations (Table 10) proved advantageous in the Konkan region (Nagwekar *et al.*, 2014). Although there was no significant variation in coconut yield during the first two years after planting the tree spices, significant variations were observed from the third year onwards. Among the blocks, the nutmeg block produced the highest yield (121 nuts per palm), followed by the cinnamon block (114.22 nuts per palm), while the coconut mono-crop block had the lowest yield (78 nuts per palm).

Table 9: Average coconut yield per palm both before and after planting spice crops, along with the percentage increase observed in various blocks of spice crops [Source: Parulekar *et al.*, 2020]

Block particulars	Block						
	Black Pepper	**Nutmeg**	**Cinnamon**	**Garcinia**	**All spice**	**Clove**	**Control**
Average yield before planting spices	83	71	69	64	49	47	77 (Average yield of 26 years)
Average yield after planting spices 1982 to 2003	102	119	118	96	93	92	
Percent increase	23	69	71	50	90	96	-

Fig. 5: Intercropping of nutmeg, black pepper and cinnamon with coconut [*Source*: Parulekar *et al.*, 2020]

Table 10: Economics of Coconut-based high density multi-species cropping system with average of 5 years/ha: 1999-2004 [Source: Parulekar *et al.*, 2020]

Block	Total Cost (₹)	Total Returns (₹)	Net Profit (₹)	B:C Ratio
Black Pepper	79313	130213	50900	1.64
Nutmeg	86417	179995	93578	2.08
Cinnamon	83449	137877	54428	1.65
Garcinia	81483	136976	55493	1.68
All spice	68087	105952	37865	1.55
Clove	77017	109496	32479	1.42
Control	55207	81194	25987	1.47

14. Shri Jaykishan Ramdas Shinde, from Varudi village in Jalna District, Maharashtra, has developed a Horticulture-based Integrated Farming System (IFS) model on his 1.30 hectare farm. The farm consists of various enterprises, including pomegranate cultivation (0.60 hectares with 564 plants), sericulture with mulberry plants (0.40 hectares with 5500 plants and a 0.02 hectare sericulture shade), a goat shed (0.02 hectares for 20 goats), fisheries with a farm pond (0.20 hectares and 4000 fish), poultry (0.02 hectares for 100 birds), a farm house (0.02 hectares), five animals (0.02 hectares), and the cultivation of vegetables and pulses using organic farming practices. Over the past three years, the IFS has generated an average net profit of ₹633000/- per annum from the entire 1.30 hectare area (equivalent to ₹4.71 lakhs per hectare). This successful model has been replicated among 80 farmers in the state, with 35 of them adopting it in the Jalna district alone.

15. The study was conducted in ten villages located in two tribal-dominated blocks, namely Aburoad (Jhamar, Fatehpura, Moongthala, Dhamsara and Chandella) and Pindwara (Amli, Rajpura Kerlapadar, Moori, and Talabfali) in Sirohi district (Rajasthan) between 2008-09 and 2012-13. Based on the available resources, various horticulture-based farming system models (Table 11) were implemented, proving to be economically viable for small and marginal farmers. In addition to these models, a range of 7-10 modern agro-technologies were introduced, such as micro-irrigation, hybrid seeds, paired row planting, machinery rental, off-season cultivation, vegetable staking, nutrient management, mixed cropping, grading, primary processing, packing, and retail marketing. After four years, a field survey was conducted in 2012-13 to evaluate the impact of the farming system models and modern agro-technologies.

Table 11: Model of farming system based on horticulture 2.0 hectare through crop diversification [Source: Bhardwaj and Vyas, 2015]

Sl. No.	Models	Type of model	Main components
1	Model-1	Crop production	*Kharif* – Bajra, maize, cluster bean, green gram, etc. *Rabi* – Wheat, barley, gram, mustard etc.
2	Model-2	Cash crop production	Fennel – 0.5 ha Castor – 0.5 ha Cotton – 0.5 ha Other crop – 0.5 ha
3	Model-3	Vegetable production	*Kharif* – Cucurbits, okra *Rabi* – Brinjal, tomato, pea, cole-crop etc. *Zaid* – Chilli, okra, cucurbits
4	Model-4	Crop + cash crop + vegetable production	Cash crop – 0.7 ha Crop – 0.6 ha Vegetable – 0.7 ha
5	Model-5	Crop + cash crop + vegetable + fruit production	Cash crop – 0.5 ha Crop – 0.5 ha Fruit plants – 0.5ha Vegetable – 0.5 ha
6	Model-6	Crop + cash crop + vegetable + flower + fruit production	Cash crop – 0.4 ha Crop – 0.4 ha Vegetable – 0.4 ha Fruit plants – 0.4ha Flowers – 0.4 ha
7	Model-7	Crop + cash crop + vegetable + flower + fruit production with retail marketing	Cash crop – 0.4 ha Crop – 0.4 ha Flower – 0.4 ha with retail marketing Fruit production – 0.4 ha Vegetable – 0.4 ha

The agro-ecological conditions in the area were suitable for diversifying traditional agriculture into more profitable farming practices. The data presented in Table 12 showed that the farmers who adopted Model-7 (crop + cash crop + vegetable + flower + fruit production with retail marketing) earned ₹5.30 lakh per year with the highest benefit-cost ratio (4.38) and generated 1800 man-days of employment opportunities from 2 hectares of cultivable land. In contrast, farmers using Model-1 (crop production) earned ₹1.22 lakh per year with the lowest benefit-cost ratio (2.03) and created a minimum number of employment opportunities (480 man-days). Previous studies have shown that horticulture-based integrated farming systems are more profitable and effectively utilize available resources to generate on-farm employment opportunities. The cultivation of tomatoes and okra has significantly improved the socio-economic conditions of the farming community. During the year 2012-13, farmers were earning ₹40,000 to ₹1 lakh from just 0.25 hectares of land, whereas previously they earned only ₹5,000 to ₹8,000 from the same plot.

Table 12: Table-XII: Calculations of economics along with employment generation under various Horticulture-based farming system models [Source: Bhardwaj and Vyas, 2015]

Models	Cost of Cultivation (₹ in lakh)	Gross Return (₹ in lakh)	Net Return (₹ in lakh)	Benefit: Cost Ratio (B:C ratio)	Employment Generated
Model-1	0.60	1.82	1.22	2.03	480 Man-days
Model-2	0.78	2.85	2.06	2.64	780 Man-days
Model-3	1.68	6.80	4.12	3.04	1250 Man-days
Model-4	1.02	2.85	2.91	2.86	850 Man-days
Model-5	0.90	4.32	3.33	3.75	1170 Man-days
Model-6	1.05	5.07	4.01	3.80	1500 Man-days
Model-7	1.20	6.50	5.29	4.38	1800 Man-days

References

Anonymous. (2008). Complete Package and Practices for the Cultivation of Vegetable Crops in Himachal Pradesh. Directorate of Extension Education Pub, CSK HPKV, Palampur, Himachal Pradesh, pp. 203.

Bahera, U.K. and Mahapatra, I.C. (1998). Income and employment generation for small and medium farmers through integrated farming system. Indian Farming, 48: 16-28.

Bhardwaj, R.L. and Vyas, L. (2015). Development of Region Specific Horticulture based Integrated Farming System Models with Crop Diversification for Sustainable Livelihoods and Nutritional Security of Tribal of Sirohi District of Rajasthan: A Case Study. Advances in Social Research, 1(1): 17-29.

Bhooshan, N. and Sharma, A. (2021). Avenues of agri-business entrepreneurship in India. Indian Farming, 71(9): 40-43.

Channabasavanna, A.S., Biradar, D.P., Prabhudev, K.N., and Hegde, M. (2009). Development of profitable IFS model for small and medium farmers of Tungbhadra project area of Karnataka. Karnataka Journal of Agricultural Sciences, 22(1): 25-27.

Choudhary, A.K., Thakur, S.K. and Yadav, D.S. (2012). Development of integrated farming system model for marginal and small farmers of Mandi district of Himachal Pradesh- an innovative extension tool. Journal of Hill Agriculture, 3(1): 46-52.

Gangaiah, B., Bohra, P., Haridas, H., Bala, P.A., Saktivel, K., Ahmed, S.K.Z., Bommayasamy, N., Saravanan, K., Singh, S.K., and Swarnam, T. P. (2019). An Overview of Integrated Farming Systems of Coastal India. ICAR- Central Island Agricultural Research Institute, Port Blair, pp. 1-185.

Kashyap, P., Kansal, A., Prusty, A., and Singh, J. (2015). Evaluation of Horticulture Based IFS Models for Providing Nutritional Security to Small and Marginal Farmers of Western Plain Zone of Uttar Pradesh, India. International Journal of Economic Plants, 2(1): 015-017.

Kerutagi, M., Talavar, M., and AS, P. (2019). Impact of horticulture based integrated farming system on farmer's income and welfare in Northern Karnataka. Journal of Pharmacognosy and Phytochemistry, 8(3): 1010-1019.

Kumar, R., Patra, M., Thirugnanavel, A., Deka, B., Chatterjee, D. and Borah, T. (2018). Comparative evaluation of different integrated farming system models for small and marginal farmers under the Eastern Himalayas. Indian Journal of Agricultural Sciences, 88(11): 1722-1729.

Nagwekar, D.D., Haldankar, P.M., Arulraj, S., and Maheshwarappa, H.P. (2014). Lakhi baug for realizing maximum income from coconut. Indian Coconut Journal, 57: 23-26.

Parulekar, Y. R., Khandekar, R. G., Haldankar, P. M., Babu, N., Cheriyan, H., Pethe, U.B., Thorat, S.B., Dalvi, N. V., Kulkarni, M. M., Gawankar, M. S., Haldavnekar, P. C., Pawar, C. D., Mali, P. C., and Rathod, R. R. (2020). Integrated Multi-storeyed Cropping of Tree Spices in Coconut Plantations for Sustainability under Aberrant Coastal Climate. Advanced Agricultural Research & Technology Journal, IV (1): 105-113.

Patil, J.L., Haldankar, P.M., Jamdagni, B.M. and Salvi, M.J. (1991). Influence of intercropping with tree spices on yield of coconut (Cocos nucifera L.). Indian Coconut Journal, 22: 15-18.

Rani, D.J. (2015). Success through farm integration- A case study of a farmer at Dhauli in the outskirts of Bhubaneswar City in Odisha. IOSR Journal of Agriculture and Veterinary Science (IOSR-JAVS), 8(1): 51-53.

Roy, R. and Kharga, B. (2022). Integrated Farming System: A boon in hill farming system. Indian Farming, 72(1): 23-26.

Sahoo, M.R., Roy, S.S., Prakash, N., and Ngachan, S.V. Horticulture based farming system for sustainability and higher profitability. A Comprehensive Manual on Integrated Farming System: An Approach towards Livelihood Security and Natural Resource Conservation. Edited by Prakash, N., Roy, S.S., Ansari, M.A., and Sharma, S.K., ICAR Research Complex for NEH Region, 2015, pp. 46-54.

Sambrani, V. (2016). Government Policies and Rural Entrepreneurship – Case Study of a Horticulture Entrepreneur. Journal of Entrepreneurship and Management, 5(2): 34-41.

Singh, A., Tanwar, S.P.S., Meghwal, P.R., Kumar, P., and Kumar, P. (2017). Exploring horticulture-based integrated farming for more return. Indian Horticulture, 62(5): 3-7.

Sudha, B., John, J., Meera, A., Sajeena, A., Jacob, D., and Bindhu, J. (2021). Coconut based integrated farming: A climate-smart model for food security and economic prosperity. Journal of Plantation Crops, 49(2): 104-110.

Techno Serve (2018). Integrated Farming for Improved Futures in India [Photograph]. Web: https://www.technoserve.org/blog/integrated-farming-for-improved-futures-in-india/ [Accessed on 16th July, 2023] [illegible] (2013) [illegible] integrated farming system model [illegible] development [illegible] *Journal of [illegible]* [illegible]

Kumar, [illegible] (2012) [illegible] *Integrated Farming System* [illegible]

[illegible] (2011) [illegible] systems [illegible] integrated farming systems [illegible]

[illegible], M. [illegible] (2012) [illegible] impact of [illegible] system [illegible] and [illegible] *and Biotechnology* [illegible]

[illegible] (2010) [illegible] comparative evaluation [illegible] farming [illegible]

[illegible] (2013) [illegible] integrated farming [illegible]

Panwar, [illegible] (2012) [illegible] integrated farming system [illegible]

[illegible] (2007) [illegible] *Journal* [illegible]

[illegible] (2012) [illegible] integrated [illegible]

Rao, [illegible] (2012) [illegible] *Journal* [illegible]

[illegible] integrated farming system [illegible] production [illegible]

Singh, [illegible] (2006) [illegible] integrated farming [illegible]

[illegible] (2011) [illegible] integrated [illegible] *Journal of* [illegible]

[illegible] (2011) [illegible] Bulletin [illegible] India [illegible]

[illegible] (2014) [illegible] integrated farming system [illegible]

3

Horticulture-based Integrated Farming Systems An Emerging Trends of Enterprise

Harjot Singh Sohi and P. S. Tanwar

Krishi Vigyan Kendra, Barnala, Guru Angad Dev Veterinary and Animal Sciences University, Ludhiana – 141004, Punjab

Abstract

Horticulture-based Integrated Farming System (IFS) must adopt appropriate soil, water crop, and pest management techniques that are both economically and ecologically responsible. IFS based on horticultural crops make better use of resources by converting waste from one organisation into input for another. Horticulture wastes may be utilised as animal feed, while livestock manure can increase horticultural crop output. By boosting the soil's levels of nitrogen, phosphorus, organic carbon, and microbial life, IFS also contributes significantly to enhancing soil health and lowering the need for artificial fertilizers. When beekeeping is systematically implemented as an adjunct to farming, it can help communities grow crops by improving pollination of the crops and vice versa horticulture crops provide flora for bees. Many farmers were able to learn about honeybee production and its significance in IFS based on horticultural crops. An integrated horticulture-aquaculture farming strategy results in better resource management and higher profits. Comparing this integrated system to aquaculture alone, the return was 20–25% greater. Additionally, mushroom compost or wasted mushroom substrate is utilised as a biofertiliser and may be employed in a number of integrated farm management practices for horticultural crops. This study collates information on how to potentially include different components in horticultural-based integrated farming systems as a better source of income.

Keywords: IFS, Horticulture, Dairy, Bee keeping, Fishery and Mushroom.

Introduction

The ongoing need to create effective agricultural systems that can meet a farmer's needs is driven by the demand for food, feed, fodder, and natural resources. More than 86 percent of Indian farm families, compromising marginal and small farmers, live in risky, diversified production conditions. There is limited room for horizontal growth of land for agriculture due to the increasing population and declining land per capita. Small and marginal farmers, who make up 84.97 percent of all operational holdings and operating 44.31 percent of the area, need a powerful tool to ensure income, livelihood security, nutritional security and employment in a sustainable manner. This tool is known as the Integrated Farming System (IFS) (Pawar *et al.*, 2019). Integrated horticulture farming is a term used when horticulture become a major commodity. Consequently, integrated farming is the technique of mixing horticulture crops with agronomy or livestock to fully utilise resources and boost production. In addition to contributing to the health, happiness, and wealth of the populace, the cultivation of fruit crops under the horticulture-based farming system (HBFS) also ensures sustainable farm income despite harsh climatic conditions. The integration of horticultural modules into farming systems helps increase productivity and profitability, especially when growing vegetables (Kashyap *et al.*, 2015). To increase productivity, profitability, and nutrient recycling as a viable option for tripling farmers' income, the ICAR-Indian Institute of Farming Systems Research (IIFSR) in collaboration with various State Agricultural Universities developed 26 scientifically designed IFS models with horticulture components belonging to 14 NARP zones covering 20 states (DARE, 2014-15). India now produces more horticultural goods (320.48 million tonnes) than food grains. In comparison to food grain production, horticulture crops produce 12.49 tonnes per hectare as opposed to 2.23 tonnes per hectare. India, which is also the world's second-largest producer of fruits and vegetables, has emerged as a leader in the production of several fruits, including mango, guava, papaya, sapota, banana, pomegranate, lime, and aonla.

Goals of the Horticulture-based Integrated Farming System

- Sustainable development of rural communities' farm-hold systems.
- Better input efficiency for the enhancement of the agricultural production system.
- Meeting the nutritional demands of farm households while also providing for their basic needs.

- Increasing the family's income through efficient resource usage and effective systemic recycling.

Types of Horticulture-based Integrated Farming System

1) Horticulture – Dairy
2) Horticulture – Bee keeping
3) Horticulture – Fishery
4) Horticulture – Mushroom
5) Horticulture – Goat

Basically, the "horticulture-based integrated farming systems" depend on the cropping pattern of the particular area under study. Different types of agricultural entrepreneurship are interconnected to create an IFS model.

1) Horticulture – Dairy

The horticulture-livestock integrated farming system or the integration system of dairy and plants generally has not been used much. When comparing different integrated farming systems as a whole, the average net returns were higher (₹99539/-) than with a non-integrated agricultural system (₹19098/-). However, in comparison to a non-integrated farming system, integrated horticulture-based farming was a successful endeavour and had a favourable impact on the lifestyle and financial standing of the agricultural community (Kerutagi *et al.*, 2019).

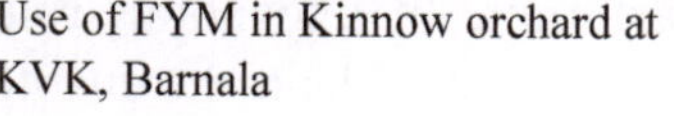

Use of FYM in Kinnow orchard at KVK, Barnala

Feeding of pea husk to dairy animals at village Manpur, Fatehgarh Sahib, Punjab

Fig. 1: Integrated Horticulture + Dairy Farming

About 7.7% of the crude protein and 5.0% of the ether extract are present in dried apple pomace. For nursing dairy cows, it provides 1.86 Mcal metabolisable energy (ME)/kg DM and 1.06–1.12 Mcal net energy (NE)/kg DM (NRC, 2001). Due to its excellent OM digestibility (85–90%) and energy availability

(2.76–2.9 Mcal ME/kg DM and 1.66–1.76 Mcal NE/kg DM) for nursing dairy cows, dried citrus pulp is used as a grain alternative in concentrate diets. The availability of ME is 85% to 90% of maize's and equivalent to barley's (NRC, 2001; Bampidis and Robinson, 2006). Ground field peas can make up as much as 40% to 50% of beginning diet concentration taking the place of some of the maize, barley, or soybean meal. In experiments in Minnesota and Alberta, similar animal performance was noted for developing heifers, as field peas can serve as their entire supply of protein (Anderson *et al*., 2002) and a large number of dairy farmers in Punjab commonly use pea husk in the feeding of cattles. In Figure 1, the majority of wastes from fruits and vegetables are highly fermentable and perishable due to their high moisture (80–90%), total soluble sugars (6–64%), and crude protein (10–24%) contents. A huge number of these resources are accessible at the busiest times of production or processing, and because they cannot be used up as quickly as they become available, they accumulate and potentially pollute the environment. Therefore, appropriate measures should be taken to preserve these resources so they can be fed to the animals year-round or, more particularly, during the lean season (Wadhwa and Bakshi, 2013).

Similarly, among bulky organic manures, the usage of farm yard manure (FYM) has a significant place. FYM appears to directly affect crop output by accelerating the respiration process through increased cell permeability or by hormone growth action. Through biological decay, it provides plants with N, P, and K in forms that they can use (Figure 1). It has been observed that higher fruit production (27.62 t ha^{-1}) in strawberries can be achieved by applying 40 tonnes of organic fertiliser (FYM) with 60 kg per hectare of NPK fertilisers (Mahadeen, 2009).

2) Horticulture – Bee Keeping

As part of integrated agricultural practises, farmers and landless labourers in rural areas engage in the agro-based activity of beekeeping. Beekeeping enhances the capacity of rural populations to obtain employment, earn income, and satisfy their nutritional needs. Honeybees perform a vital economic role in nature by pollinating enormous numbers of flowering plants to enable the development of seeds or fruit, despite the fact that they are most well-known for producing honey. Honeybees have been contributing to civilisation by guaranteeing the pollination of cross-pollinated crops (Figure 2) and by offering honey and a variety of other beehive products. A healthy environment depends on honey bees to preserve plant variety. Pollination by honey bees also improves product quality.The possible advantages of bee pollination in terms of an increase in agricultural production range from 5% to 33150% (Sarswat, 2017) in Table 1.

Table 1: The following list of horticultural crops and the boost in minimum productivity brought on by bee pollination

Sl. No.	Fruit Crops	% Increase in Yield	Vegetables Crops	% Increase in Yield
1	Apple varieties	180	Radish	22
2	Pears	240	Cabbage	100
3	Plums	6.7	Turnip	100
4	Cherry	56.1	Carrot	9.1
5	Strawberry	17.4	Onion	353.5
6	Raspberry	291.3	Brinjal	35 to 67
7	Persimmon	20.8	Cucumbers	21.1

Bee-keeping at KVK, Barnala

Pollination and collection of nectar by honey bees at KVK, Barnala

Fig. 2: Integrated Horticulture + Bee-keeping

In light of the aforementioned, honey bees and beekeeping have emerged as a 5th input for agriculture, which governs the efficacy of the other four inputs. These 4 inputs include land, labour, capital, and management, which includes seed, fertiliser, pesticides, water, machinery, etc. In an investigation carried out in Punjab between 1982 and 1987 (Chahal *et al.,* 1988) revealed that honey bees obtain their nourishment from the blossoms of 172 plant species, of which 27 plant species are primarily used for nectar, 25 plant species are used for pollen, and 120 plant species offer both nectar and pollen as shown in Figure 2. In order to promote their health and raise the productivity of honey bees, pollen supplies protein, while nectar serves as a source of sugar. Bee farming could only develop in locations with the largest density of bee flora. Numerous studies have demonstrated that bees' pollination services significantly increase apple fruit output (Sharma *et al.*, 2012).

3) Horticulture – Fishery

In an integrated fishery-based farming system, the top, inner, and outer dykes of ponds as well as nearby lands are often used for horticulture plants including floriculture, olericulture (fruit and vegetables), agroforestry, etc. In order to increase the efficiency of the fish production system, horticulture integration with fish produces 20-25% more output (Bhuyan and Neog, 2018). In Figure 4, pond water is utilised for irrigation, and crops, vegetables, and fruit-bearing plants are grown using silt, a high-quality manure. In this integrated system, horticulture is the secondary crop, hence extreme caution should be used while choosing crop varieties to avoid impeding fish production. Selection criteria for horticulture crops include –

- The embankments must not be damaged by crops or eroded.
- Crops cannot have a deleterious impact on pond water.
- Crops should ideally be applicable to fish culture.
- Crops raised must be compatible with the local ecology and soil.

Additionally, plants grown for vegetables and fruits should be smaller, less shaded, evergreen, seasonal, and profitable. The dwarf varieties of mango, litchi, guava, banana, papaya, and lemon are suitable fruit plants for this use. Plants like pineapple, ginger, turmeric, and areca nuts are used as intercrops on the part of the pond dike that is planted with coconut trees. To recycle the green leaves of farmed crops, herbivorous fish raised in this method are best. In the central Brahmaputra valley of Assam, fish are being cultured, including grass carp (*Ctenopharyngodon idella*) and java barb (*Puntious japonicus*). By combining fish species like grass carp and IMCs with horticulture, farmers in this region have achieved an output of over 2500 kg/ha. Residues of vegetables and green fodder can easily be recycled into fish ponds and herbivorous fish like grass carp (monoculture) at a stocking density of 5000/ha has resulted in 3,500-5,000 kg/ha production. In mixed culture practice of grass carp, rohu, catla, mrigal, silver carp and common carp in a 50:15:15:10:5:5 ratio at a density of 7,500/ha yielded 4,000-5,000 kg of fish fed with dyke-grown green fodder (Ayyappan *et al.*, 2011). The addition of horticulture to fish farming increases productivity by 20-25%, making it an efficient way to increase fish production and also generates engagement opportunities throughout the year. Additionally, integrated agricultural practises with aquaculture utilises the non-edible parts of plants (Tewari and Kaur, 2022), recycle the nutrients and water inputs and have superior effects on the production cycle (Kaur *et al.*, 2023). Integration offers a wide range of social and ecological advantages, but it may also be successful on a modest semi-subsistence scale. Weak integration brought on by issues including nutritional deficiencies, overuse of agrochemicals, and a lack of knowledge and management skills might limit this system's performance. Therefore, a new adaptable and comprehensive approach is required to spread awareness of this system.

Use of fish pond water in horticulture crops at village Alkara, Barnala

Use of fish pond dyke for horticultural crops at village Alkara, Barnala

Fig. 3: Integrated Horticulture + Fish Farming

Horticulture – Mushroom

Following crop harvest, the agricultural/horticultural residue produced from the cropland is typically burnt or destroyed in the crop field, posing environmental issues. As a result, this waste is alternatively handled by being used as a substrate for the growth of mushrooms. According to (Sharma *et al.*, 2017), the Indian mushroom industry generated revenue of 7282.26 lacs by exporting 1054 quintals of white button mushrooms, which account for 73% of the country's total production of mushrooms, followed by oyster mushrooms (16%), paddy straw mushrooms (7%) and milky mushrooms (3%). Although agricultural waste generated from farming can be used in a variety of agro-based applications and other industrial applications, the cost of collection and transportation is higher than the revenue it generates by using such agricultural residues (Bhuvaneshwari *et al.*, 2019). However, it could be used advantageously if a mushroom farm were to be established on the farm's site, so mushrooms would be good additions to the agricultural system.

Similarly, SMC (spent mushroom compost) (Figure 4) is the leftover compost waste produced by or left outside after the manufacture of mushrooms. This compost is easily accessible, and its typical composition includes a combination of decomposed wheat straw, horse manure, crushed chalk, etc. The SMC is an ideal habitat for various microbes, including bacteria and fungi, due to its abundance of salt and organic material as well as enzymes and other nutrients. Because these microbes work together to suppress disease and promote plant growth, an integrated management strategy could be used to control the severity of the pathogens. Composts' ability to suppress disease depends on a number of variables, including microbial activity, microbial population dynamics, nutrient concentrations, and other related chemical and physical variables (Patil *et al.*, 2018). Different studies suggest that SMC could be used as a source of biocontrol agents against various plant diseases due to its abilities to produce organic salts and enzymes, and it has been reported that SMC has antagonistic fungi (*Trichoderma* sp.) and bacteria, SMC not only controls the disease but also promotes plant growth. According to (Fidanza *et al*., 2010), fresh spent mushrooms typically have a C:N ratio of 13:1. The nutrients in SMS and SMC depend on the species of mushroom from which the spent material was derived as well as the soil type where mushrooms are grown. Horticultural conditions, post-crop management, the age of the spent material, its storage time, and post-crop composting have a significant impact on SMC and SMS characteristics. As soil conditioners, SMS and SMC have similar functions in the soil, such as enhancing soil structure, supplying nutrients to plants, increasing the availability of those nutrients, increasing soil microbial population, enhancing soil cation exchange capacity, enhancing plant root structure, enhancing soil aeration, enhancing soil water status, reducing soil compaction, and regulating soil pH. As a result, SMC and SMS are appropriate for usage in horticulture. The ARYA initiative, ASCI's training programme for mushroom growers, and KVK Barnala's vocational training programmes have all been determined to be crucial in helping rural adolescents, farmers in using the mushroom spent compost and other horticulture techniques (Sohi *et al*., 2021).

Button mushroom at KVK, Barnala

Use of spent mushroom waste in orchards

Fig. 4: Integrated Horticulture + Mushroom

Similarly, SMC (spent mushroom compost) (Figure 4) is the leftover compost waste produced by or left outside after the manufacture of mushrooms. This compost is easily accessible, and its typical composition includes a combination of decomposed wheat straw, horse manure, crushed chalk, etc. The SMC is an ideal habitat for various microbes, including bacteria and fungi, due to its abundance of salt and organic material as well as enzymes and other nutrients. Because these microbes work together to suppress disease and promote plant growth, an integrated management strategy could be used to control the severity of the pathogens. Composts' ability to suppress disease depends on a number of variables, including microbial activity, microbial population dynamics, nutrient concentrations, and other related chemical and physical variables (Patil *et al*., 2018). Different studies suggest that SMC could be used as a source of bio control agents against various plant diseases due to its abilities to produce organic salts and enzymes, and it has been reported that SMC has antagonistic fungi (*Trichoderma* sp.) and bacteria, SMC not only control the disease but also promote plant growth. According to (Fidanza *et al*., 2010), fresh spent mushrooms typically have a carbon:nitrogen ratio of 13:1. The nutrients in SMS and SMC depend on the species of mushroom from which the spent material was derived as well as the soil type where mushrooms are grown. Horticultural conditions, post-crop management, the age of the spent material, its storage time, and post-crop composting have a significant impact on SMC and SMS characteristics. As soil conditioners, SMS and SMC have similar functions in the soil, such as enhancing soil structure, supplying nutrients to plants, increasing the availability of those nutrients, increasing soil microbial populations, enhancing soil cation exchange capacity, enhancing plant root structure, enhancing soil aeration, enhancing soil water status, reducing soil compaction, and regulating soil pH. As a result, SMC and SMS are appropriate for usage in horticulture. The ARYA initiative, ASCI's training programme for mushroom growers, and KVK Barnala's vocational training programmes have all been determined to be crucial in helping rural adolescents, farmers in using the mushroom spent compost and other horticulture techniques (Sohi *et al*., 2021).

4) Horticulture – Goat

The goat industry in India generates 8.4% of the country's 38,590 crore livestock GDP, or ₹22625 crores from meat, ₹9564 crores from milk, ₹1491 crores from skin, ₹1535 crores from manure, and ₹3360 crores from other sources. The Indian goat industry also contributes 4.2% of rural jobs to small, marginal, and landless farmers. India produced 5,53,380 tonnes of goat meat in 2019, which placed it as the second-largest producer in the world. In 2020, India was the fifth-largest exporter of goats (live), with a value of about 6.17 million dollars. Goat milk production in India rated as the top in the world in 2017, with an estimated 50 lakh metric tonnes produced (Saravanan, 2022). The Indian government is currently promoting integrated farming as a sustainable agricultural venture for both rural and urban enterprises. Goat raising is a crucial part of integrated farming. The value of goat farming in integrated farming is increased by the use of goat manure in horticulture, the ability to control vegetation by raising goats, and the ability to generate high income through goat rearing with no inputs. Goats are meticulous creatures. They dislike silage and old straw. They enjoy green fresh legumes much. They are hesitant to accept the feeders' stale, dropped food. As a result, feed should be provided to them in split dosages three to four times each day. In stall feeding (Figure 5), green forages are also provided two to three times daily. Goats like foraging on common green plants, such as lucerne, berseem, napier grass, and cowpea. Goats may be fed tree forage such as subabul, tamarind, glyceridia, and peepal.

Two-thirds of forages with tree fodder are favoured. Unwanted leaves that are left over after harvesting stems or roots of plants like kohrabi, turnips, parsnips, and sweet potatoes. Unsellable, pest-infested greens like kale, lettuce, and cabbage that are damaged yet still palatable to goats. Plant residue from production, including broccoli, Brussels sprouts, peas, beans, and cowpeas (Reuter, 2015). Compared to cow and buffalo dung, goat manure has 2.5 times more nitrogen and phosphorus. Farmers in agriculture and horticulture thus place more trust in goat manure. The annual dung output of a typical goat is around 300 pounds, and the feces are in pellet form, making them simple to manage. Potassium, nitrogen, and potentially other minerals are all abundant in goat dung (Osuhor, 2002). A single individual may raise 30-40 goats while using a grazing system, and 40-80 goats while using stall feeding. Compared to growing other animals, goat farming requires less labour (Kattimani, 2021).

Goat farm unit at KVK, Barnala

Use of goat manure in horticultural crops

Fig. 5: Integrated Horticulture + Goat Farming

Future Aspects of Horticulture-based Integrated Farming System

a) **Recycling of Waste:** Utilise waste from one component at the lowest possible cost to be profitable. As a result, production costs are reduced, waste materials are better used, and middlemen are removed from the majority of input processes. Calculating net profit increases the B/C ratio.

b) **Potentiality or Sustainability:** By using the byproducts of connected components effectively, organic supplementation offers the chance to sustain the production base's potentiality for considerably longer durations.

c) **Balanced Food:** We combine ingredients of various types to provide various sources of nutrients.

d) **Environmental Safety:** By combining the right components, waste materials are efficiently recycled in Horticulture-based: 'Integrated Farming System', reducing environmental contamination.

e) **Income Generation:** Due to business interactions with crops, dairy, bee keeping, honey and mushrooms income rounds out the year providing farmers with a constant source of funding.

Conclusion

IFS is a multidisciplinary whole farm strategy that is excellent at addressing the issues faced by small and marginal farmers. The strategy involves merging different farm components and recycling agricultural leftovers and byproducts on the farm itself to increase revenue and employment from small holdings. Protective food and environmental safety ensure increased production, profitability, and sustainability. Other significant advantages of a horticulture-based: 'integrated farming system' include the recycling of waste, year-round revenue, energy savings, addressing the fodder shortage, job creation, and ultimately an increase in the standard of living for farmers.

References

Anderson, V., Harrold, R., Landblom, D., Lardy, G., Schatz, B. and Schroeder, J.W. (2002). Book: A Guide to Feeding Field Peas to Livestock. North Dakota State University.

Ayyappan, S., Moza, U., Gopalakrishan, A., Meenakumari, B., Jena, J.K. and Pandey, A.K. (2011). Handbook of fisheries and aquaculture (2nd ed.). Directorate of Information and Publications on Agriculture, Indian Council of Agricultural Research.

Bampidis, V.A. and Robinson, P.H. (2006). Citrus by-products as ruminant feeds: A review. Animal Feed Science and Technology, 128: 175–217.

Bhuvaneshwari, S., Hettiarachchi, H. and Meegoda, J.N. (2019). Crop residue burning in India: Policy challenges and potential solutions. International Journal of Environmental Research and Public Health. 16(5): https://doi.org/10.3390/ijerph16050832

Bhuyan, S. and Neog, M. (2018). "Integrated Horticulture-Fish Farming System: An approach towards sustainable development and nutritional security" Compendium on Model training course on Integrated Farming System for Sustainable Agriculture Publisher: Directorate of Extension Education, AAU, Jorhat-13.

Chahal, B.S., Brar, H.S., Goryiana, G.S. and Jhajj, H.S. (1988). Book: Italian honey bee management system. ISBN: 978-93-86267-54-2.

Fidanza, M.A., Sanford, D.L., Beyer, D.M. and Aurentz, D.J. (2010). Analysis of fresh mushroom compost. Hort Technology. 20: 449-53.

Kashyap, Poonam, Kansal, A., Prusty, A.K. and Singh, J.P. (2015). Evaluation of horticulture-based IFS model for providing nutritional security to small and marginal farmers of western plain zone of Uttar Pradesh. Int. J. Ecol. Pranks, 2(1): 15-17.

Kattimani, T.S. (2021). Advantages and challenges in goat farming. Sheep goat husbandry.

Kaur, R., Tanwar, P.S. and Shah, T.K. (2023). Common practiced integrated fish farming systems in the northern states of India. Chapter 15 of Entrepreneurship in Integrated Farming System.

Kerutagi, M.G., Talavar, M. and Pavitra, A.S. (2019). Impact of horticulture based integrated farming system on farmer's income and welfare in Northern Karnataka. Journal of Pharmacognosy and Phytochemistry. 8(3): 1010-19.

Mahadeen, A.Y. (2009). Influence of organic and chemical fertilisation on fruit yield and quality of plastic house grown strawberry. Jordan J. Agric. Sci. 5(2):167- 77.

NRC. (2001). Nutrient Requirements of Dairy Cattle. 7th revised edn. National Research Council, National Academy of Sciences, Washington, D.C., U.S.A.

Osuhor, C.U. (2002). "Research Note: Manure Production by Goats Grazing Native Pasture in Nigeria." Tropical Grasslands. 36: 123-125.

Patil, M.G., Rathod, P.K. and Patil, V.D. (2018). Compost : A tool for Managing Soil Borne Plant Pathogens. International Journal of Current Microbiology and Applied Sciences. 6: 272-280.

Pawar, A.S., Kashyap, P., Ravisankar, N., and Prusty, A.K. (2019). Horticulture-based Integrated Farming System: A viable option for doubling farmers' income. Conference paper.

Reuter, E. and Chert Hollow Farm (2015). Feeding Greens and Other Garden Materials to Dairy Goats. Homesteading and Livestock. Mother Earth News. https://www.motherearthnews.com/homesteading-and-livestock/feeding-greens-to-goats-zbcz1510/.

Saravanan, M.S. (2022). Goat Farming Practices in India https://www.pashudhanpraharee.com/goat-farming-practices-in-india/

Sarswat, B.L. (2017). Beekeeping: Vital Input for Sustainable Agriculture. https://pib.gov.in/newsite/printrelease.aspx?relid=157604.

Sharma, H.K., Uma, P., Bisht, K., Tewari, P. and Phartiyal, P. (2012). Impact of honeybee pollination in enhancing the apple production in Nainital district of Uttarakhand. Advances Pollen Spore Research 30: 99-102.

Sharma, V.P., Annepu, S.K., Gautam, Y., Singh, M. and Kamal, S. (2017). Status of mushroom production in India. Mushroom Research. 26(2): 111-120.

Sohi, H.S., Tanwar, P.S. and Matharu, K.S. (2021). Constraints Perceived by Rural Youth in Adoption of Button Mushroom Cultivation and its Success in District Barnala. Journal of Community Mobilisation and Sustainable Development. 16(1): 7-13.

Tewari, G. and Kaur, R. (2022). Fish feed supplementation using non-conventional plant resources: Way to sustainable aquaculture. The Pharma Innovation Journal SP-11(5): 309-321.

Wadhwa, M. and Bakshi, M.P.S. (2013). Utilization of fruit and vegetable wastes as livestock feed and as substrates for generation of other value-added products. RAP Publication. E-ISBN 978-92-5-107632-3.

4

Women-Centric Entrepreneurship Development Horticulture Based Integrated Farming System – A Case Study of Family Farming in Bihar

Richa Kumari[1], Sudhanand Prasad Lal[2] and Kumari Asmita[3]

[1]Home Science, Lalit Narayan Mithila University, Darbhanga, Bihar
[2]Department of Extension Education, P.G. College of Agriculture
Dr. Rajendra Prasad Central Agricultural University, Pusa, Samastipur Bihar
[3]AICRP on Women in Agriculture, College of Community Science
Dr. Rajendra Prasad Central Agricultural University, Pusa Samastipur - Bihar

Abstract

Entrepreneurship is considered one of the most significant variables influencing the economic development of society as well as the growth and progress of every country. In addition to human resources, effective resource management is also essential for a nation's success. Entrepreneurship is the act of being an entrepreneur. Entrepreneurs are thought to play a key function in initiating and maintaining socioeconomic development. As women cover up approximately half of the population, their societal situations significantly have an effect on the development of a country. In India, women make up about half of the potential human resource pool for all economic activities. If this half of the population is overlooked and not given access to opportunities in the agricultural, industrial, and tertiary sectors, development will not be realised. The idea of women entrepreneurs is relatively new in India. Women are now more aware of their rights and circumstances and are working in a variety of industries. They have built prosperous commercial empires of their own. They support economic expansion and the improvement of their socioeconomic circumstances. This chapter includes an introduction, history, conception, the meaning of women entrepreneurship, integrated farming systems and a study of a woman entrepreneur in the Muzaffarpur district of Bihar.

Keywords: Entrepreneurship development, Integrated Farming System, Women-centric.

Introduction

Entrepreneurship is considered one of the most significant variables influencing the economic development of society and the growth of every country. In addition to human resources, effective resource management is also essential for a nation's success. Entrepreneurship involves the act of creating something as an entrepreneur. Entrepreneurs are thought to play a key function in initiating and maintaining socioeconomic development. As women cover up approximately half of the population their societal status significantly significantly impact the country's development. In India, women represent about half of the potential human resource pool for all financial activity. If this segment of the population is overlooked and not given access to opportunities in the agricultural, industrial, and tertiary sectors, development will not be realised. The concept of women entrepreneurs is relatively new in India. Women are now more aware of their rights and circumstances and are working in a variety of industries. They have built prosperous commercial empires of their own. They support economic expansion and the improvement of their socio-economic circumstances.

History of Entrepreneurship

The history of entrepreneurship in India reflects the traditions, ethnicity and civilisation of Indian citizens. Richard Cantillon, who coined the term "entrepreneur," first recognised the entrepreneur and their risk-taking role in the early 18th century. He argued that the business owner purchases factor services at a given price with the intention of later selling their goods at that same price. A few decades later, J.B. Say provided a more comprehensive description of the entrepreneurial function, emphasising the coordination of production variables, the supply of ongoing management, as well as risk-taking. Then, J.S. Mill, taking a cue from Cantillon and Say, highlighted the difference between the entrepreneurial function and that of wealth, but Marshall upheld the Smith-Recardian custom of treating turnover as a single undifferentiated earnings-flow, relegating entrepreneurship to management – a particular category of skilled labour, negating the usefulness of this distinction. Joseph Schumpeter's work in 1911, is an exception to this generalisation as he was one of the earliest to identify the significance of the entrepreneur. However, he argued that these kinds of innovators are uncommon in society and only seldom show up on the scene of progress. Therefore, economic growth spurts depend on the development of such innovative people Max Weber emphasised the idea that adopting a certain attitude inspires entrepreneurship and a creative spirit.

Concept of Entrepreneur

One who starts and sustains a commercial or economic activity is referred to as an entrepreneur. The word "entrepreneurship" originates from the French and German words "entreprendre" and "undernehmen," both of which imply to "undertake,". Women who launch, reproduce or take up a business activity are considered to be entrepreneurs (Khanka, 2002). A person who assumes and accepts the risks associated with a business arranges the various efforts or mechanisms of Production and generates novel thoughts, conceptions, and actions (Hayes, 2021). A person who starts a business and has certain skills set, capability, knowledge, understanding, and height of self-assurance & also assume the risk of operating a firm in the absence of certainty regarding future market conditions is known as an entrepreneur (Kumari and Venkatesh, 2018).

Meaning of Entrepreneurship

The financial development of any nation is greatly influenced by the success of women entrepreneurs. Women's entrepreneurship can lead to a strong involvement in the financial stability of the family and community, eliminating poverty and promoting women's empowerment, which will help to achieve the Millennium Development Goals (MDGs). Entrepreneurship refers to the broad tendency of starting a new business in society (Begum, 1993). The series of tasks carried out by an entrepreneur is called entrepreneurship.

Definition of Women Entrepreneurship

The process by which a woman or group of women start, systematise and collaborate with production activity is known as women entrepreneurship. *Women entrepreneurs are those who come up with ideas, generate the business, organise and unite the diverse production parts, run the business, take risks, and deal with the accompanying economic uncertainties (Aradhya, 2007).* Women entrepreneurs are defined by the Indian government based on the unequal representation of women in business employment in India (Ramasamy, 2009). Numerous women entrepreneurs have created jobs for themselves and their co-founders in addition to hiring people. (Kelly *et al.*, 2011). Women have played a pivotal role during the recent pandemic of COVID-19 in 2020 in relation to women entrepreneur (Sangem, 2020).

Reasons for Women who wish to become Entrepreneurs

a) To become thriftily self-sufficient

b) To set up their personal enterprise

c) To create their eccentricity in the world
d) To accomplish excellence in their work
e) To raise confidence in oneself
f) To build up jeopardy presuming capability
g) To assert identical position in the society
h) To secure better liberty and movement

Functions of Women Entrepreneur

A women entrepreneur has played a number of functions.

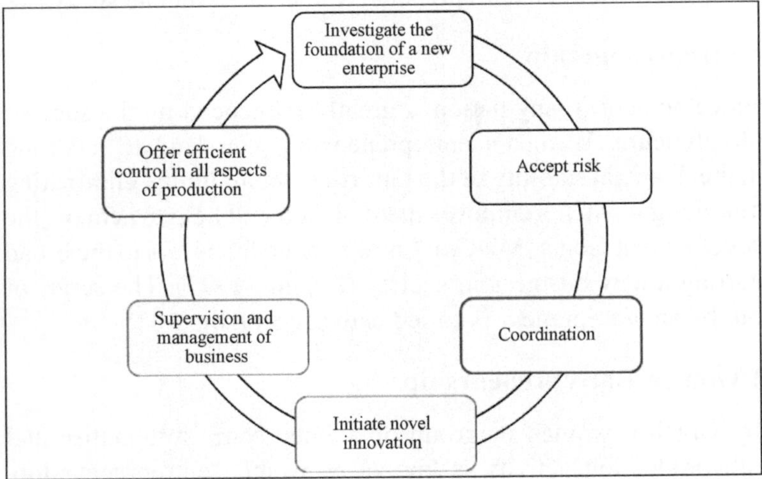

Fig. 1: Functions of women entrepreneur

Integrated Farming System

Food security can be defined as ensuring that everyone has access to enough food to meet their nutritional needs and lead active, healthy lives. The attempt to get better food security may be possible through integrated farming systems. The word "integrate" derives its roots from the Latin word "*integrare*," which means "to make whole," "to complete by adding parts," or "to merge components into a whole.

Integration often occurs when, within the framework of the farming system, the outputs (mostly by-products) of one company are used as inputs by another. Farming systems refer to farms where two or more ventures are incorporated with the farm possessions with the aim of achieving fuller consumption of existing resources for appreciating the highest earnings and also to ensure steady profits. The term "Integrated Farming System (IFS)" refers to a way of managing fish, livestock, and yields in harmony with their environments in order to prevent imminent production risks, guarantee production security, and utilise resources that are roughly compatible with the constrained amount of land, labour, and capital available. In other words, an integrated farming system (IFS), is a multifarious surrounding substance of interconnected top soil, water, plants, living things, and ecological elements, is more practical and cost-effective than the arable farming system. The major goal of the IFS is to sustain quality crop productivity and additional produce and services that supply food safety and create revenue for the rural deprived. Vegetable and horticultural crops can provide up to two to three times additional energy per unit area of land than cereal crops, and their inclusion in the current system will ensure extra nutritional security for the people. Thus, women can be involved in a variety of components of integrated farming systems including mushroom production, apiculture, aquaculture, poultry, horticulture duck rearing etc. They can also be involved in various kinds of activities in integrated farming systems *such as* irrigation, sowing and planting operations nursery management, fertiliser application, stocking, intercultural operations hoeing etc.

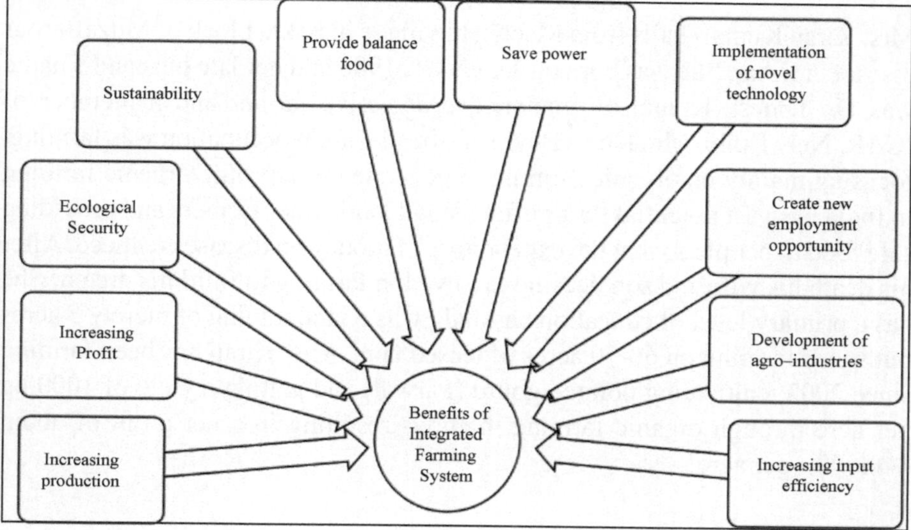

Fig. 2: Benefits of Integrated Farming System

Constraints Faced by Farmers during Applying Integrated Farming System

a) Requirement of high skills.

b) Elevated first capital venture.

c) Consequence of shadow and defoliation on yield.

d) Competition for possessions.

e) Hard to handle diverse endeavours all together.

f) Complexity in intercultural functions.

g) Complicatedness in care of animals throughout peak farming period.

h) Lengthy conversion time in tree constituent.

A Case Study of Progressive Woman Farming Family of Bihar

India is an agricultural country with 70% of its population depending on agriculture. In the past, agriculture was mainly done by males only, but in the changing environment, attitudes have shifted. Today women are also emerging as a skilled farmer. Today's family is making agriculture their main occupation. This story is about one such family whose main occupation is farming.

Mrs. Kiran Kumari hails from Machhahi village in Sakra block of Muzaffarpur district in Bihar. She was born on January 1, 1980 and her late husband's name was Dr. Dinesh Kumar, a renowned progressive farmer and a member of ICAR, New Delhi. Mrs Kiran Devi's husband's main occupation was farming, focusing mainly on organic farming to generate more profit. Organic farming in India has vast potential for uplifting small landowner farmers and providing safe food to people so that the expenditure of money on diseases reduced. After his death his wife and son became involved in farming to fulfil his dream. She has a primary level of education has and owns a landholding of merely 5 acres but she is farming on 60-70 acres of leased land. Mrs. Kiran has been farming since 2003, cultivating pointed gourd (Parwal) and getting a yield of 1000 kg per acre through organic farming methods resulting in a net profit of about ₹40,000/- per acre.

Table 1: Brief details of Mrs. Kiran Kumari

Sl. No.	Details	Specifications
1	Agro climatic zones	Middle Gangetic Plains
2	State	Bihar
3	Region	Agro- climatic zone I (Northern West)
4	District	Muzaffarpur
5	Block	Sakra
6	Village Panchayat	Machhahi
7	Village	Machhahi
8	Latitude and longitude	25.95990419795517, 85.55387223006609
9	Farmer name	Mrs. Kiran Kumari
10	Date of Birth	01-01-1980
11	Present Age	42
12	Education	Primary
13	Husband Name	Late Dr. Dinesh Kumar
14	Occupation	Organic Farming
15	Inception of farming	2003
16	Maximum output	1000 kg/acre
17	Land ownership	5 acre
18	Area under farming	Approx. 75 acre (60-70 acre on lease)

Along with parwal, she also cultivates ridge gourd (nenua), ladies' finger, brinjal, chilli, seedless lemon, paddy, wheat etc. in her field using organic fertiliser. As she uses organic fertiliser in her field for farming, the crops and vegetables produced are safe for health. Mrs. Kiran Devi looks after ploughing, sowing, harvesting, etc. in the field, while her son Sonu Kumar only 19 years old works to sell them in the market to earn a fair price for the crop. Her son Sonu Nigam Singh is cooperating with her in her farming endeavour. She also produces vermicompost on a large scale, using it in her field and selling it in the market. Her son Sonu provides training in different KVKS & ATMA farmers on farming techniques and many farmers and scientists have visited their field to see their farming techniques and produce. Her husband has received many awards from the honourable President of India, the Central Agricultural Minister, Dr. Rajendra Prasad Central Agricultural University, Pusa and ICAR, New Delhi for his commendable work in the agriculture sector. Today this family has become an inspiration for the holiday season. Now she has become a source of inspiration for all the farmers (male and female) of that locality who want to learn scientific practices to improve their livelihood. Many women of the village and their families want to make a profession by doing farming and from time to time also take proper advice related to them from Kiran Devi. She has proven that by the implementation of new technology in farming will make possible maximising the profit. This is the capital example which we have collected from field data.

Fig. 3: Mrs. Kiran Devi with their produce in the field

Fig. 4: Mr. Sonu Nigam during helping in marketing of produce

In addition to the primary field data collected for the case study of Mrs. Kiran Devi many other women entrepreneurs from Bihar have already been highlighted and have achieved great heights in her lives. Here is another example of a women entrepreneur from Bihar involved in mushroom cultivation.

Women Entrepreneurship by Mushroom Cultivation

Anita Devi belongs to Anantpur village of Nalanda district of Bihar. She graduated in home science and began producing mushrooms in 2010 after receiving training at the GB Pant University of Agriculture and Technology in Uttarakhand and the Dr. Rajendra Prasad Central Agriculture University in Samastipur, Bihar. At one point, she was in urgent need of money and sought assistance from the Krishi Vigyan Kendra in Harnaut, Nalanda. Following their advice on mushroom cultivation, she embarked on a path to success. By establishing a mushroom farm, she provided her family with a stable income and uplifted the financial status of numerous women in Anantpur and 10 neighbouring villages in the Nalanda region of Bihar by inspiring them to follow her example. Anita is famous in Bihar as the "**Mushroom Lady**" (Khan, 2017).

Conclusion

It is urgently needed to support women as they pursue their entrepreneurial dreams because women constitute a significant part of the human resource potential in the country. The economic progress of any nation depends primarily on the utilisation of human and capital resources. If we do not encourage women to participate in different entrepreneurial activities for economic improvement, the process of economic development cannot be accelerated in the country. If families, communities, and the government provide proper support to women entrepreneurs they can become an integral component of the Indian economy. Empowering female entrepreneurs is absolutely necessary for the nation to achieve its goals of sustainable development because "women have the capability to build, nurture, and transform society.

References

Aradhya, D. (2007). Women Entrepreneurship Retrieved from Your Article Library The Next Generation Library. DOI: https://www.yourarticlelibrary.com/women/women-entrepreneurship/women-entrepreneurship/99813

Begum. (1993). Entrepreneurship in Small-scale Industry: A Case Study of Engineering Units, Dhaka University. Journal of Business Studies, 14: 159.

Gill, M.S., Singh, J.P. and Gangwar, K.S. (2009). Integrated Farming System and Agriculture Sustainability. Indian Journal of Agronomy. 54(2): 128-139.

Hayes, A. (2022). Entrepreneur: What It Means to Be One and How to Get Started, Investopedia. Retrieved from https://www.investopedia.com/terms/e/entrepreneur.asp

Integrated Farming for Sustainable Agriculture and Livelihood Security to Rural Poor. (2014). Universal Researchers. DOI: https://doi.org/10.17758/iaast.a0514013

Kelly, E.L., Moen, P. and Tranby, E. (2011). Changing workplaces to reduce work-family conflict: Schedule control in a white-collar organization. American Sociological Review. 76(2): 265-290.

Khan. M.I. (2017). The Mushroom Lady of Nalanda: How this woman farmer changed the lives of hundreds of women in Bihar. DOI: https://scroll.in/article/859936/the-mushroom-lady-of-nalanda-how-this-woman-farmer-changed-the-lives-of-hundreds-of-women-in-bihar

Khanka, S.S. (2002), Entrepreneurial Development, New Delhi: S. Chand & Company. DOI: https://tinyurl.com/4z8bpwrd

Ramasamy. T. (2009). Strategic Entrepreneurial Dimensions of Self-Employed Women – A Micro Study. Southern Economist. 48(14): 39-41. DOI: http://www.shanlaxjournals.in/journals/index.php/economics/article/view/4202

Sangem, M. (2020). Challenges for women entrepreneurs in the wake of COVID-19 pandemic. Journal of Interdisciplinary Cycle Research. 12(11): 279-284.

Saraswar, R. and Lathabhavan, R. (2020). A study on Women Entrepreneurship in India. Mukt Shabd Journal. 9(6): 3422-3432.

Srivastava, M. (2009). Women Entrepreneurship in India: Some Aspects, Book of Women Entrepreneurship, Deep & Deep Publication Pvt. Ltd, pp. 1-14. DOI: https://tinyurl.com/mrfesd7c

Zade, Y.A., Toprope, V.N. and Sargar, P.R. (2021). Integrated Farming System for Sustainable Rural Livelihood Security. Just Agriculture Multidisciplinary e-Newsletter. 1(10): 1-4.

5

Tuber Crops Based Integrated Farming System for Sustainable Livelihood – A Few Case Studies

N.G. Piloo[1], N.L. Bharathi[1], N.K. Yadav[1] and Aditya Pratap Singh[2,3]

[1]College of Agriculture, Central Agricultural University, Imphal, Manipur
[2]Bidhan Chandra Krishi Viswavidyalaya, Mohanpur, Nadia, West Bengal
[3]School of Agriculture, GIET University, Gunupur, Rayagada, Odisha

Abstract

In nations like India, farming systems research is seen as a potent instrument for managing natural and human resources. With the growing population each year, it is extremely important to address the problem of supplying the basic needs of this expanding population. An integrated farming system is a multidisciplinary and efficient whole-farm strategy that might boost land productivity, farmers' income, and balance their employment, job opportunities, and sustainability needs. If effectively managed, integrated farming systems benefit from synergies among enterprises, diversity in yield, and environmental soundness, making them often less risky. Tuber crops-based farming systems are an integrated approach to farming compared to monoculture. Tuber crops may often be grown on marginal terrain with little maintenance. They require less care and can be incorporated in diverse farming systems. Most of the time, tuber crops such as cassava, lesser yam, greater yam, taro, elephant foot yam, and tannia are grown alongside vegetables, fruits, and plantation crops. In such farms, perennial produce provides financial income, while starchy roots and tubers meet some of the farm family's food needs as well as some of the animal's feed requirements. However, more than additional employment, spread of employment through-out the year has been achieved in tuber crops based integrated farming system.

***Keywords*:** Integrated farming system, Tuber crops, Sustainability, Year-round income.

Introduction

Two-thirds of the population make their living from agriculture, which also supplies raw materials for other businesses. To feed the growing population and decrease hunger and poverty, agricultural output must be increased. Various farming systems coexist in India depending on the resources available, agricultural methods, and location-specific demands of people to meet their needs for food, fuel, and fibre. An intricate, interconnected matrix is the integrated farming system (IFS) of soil, plants, animals, tools, equipment, labour, capital, and other inputs that are partially controlled by farming families, with varied degrees of political, economic, and institutional power and resting elements that affect farms. In India, the size of farm holdings has dropped over time. The demand for both basic food items and food products with added value has, however, continually increased as a result of growing income levels and population growth. By combining farming systems that give a whole farm policy and a whole systems approach to farm management, it may be possible to meet this rising demand for food with the limited resources available in India. Average farmers hardly get enough money from crop production to support their families. In addition to being a safe method of achieving reasonably high production with significant fertiliser reductions an integrated farming system approach also promotes ecological soundness and sustainable agriculture (Swaminathan, 1987). Without any explanation other agricultural practices have evolved and are currently being used by local farmers on their farms. These techniques do not make use of residues from crops, animals, or other related activities.

Increased food output, higher net income, enhanced productivity, and a reduction in the pay gap between agricultural labourers and urban industry workers are all benefits of integrating farm enterprises. Increasing non-farm employment is also crucial for raising rural residents' incomes and standards of living (Chadha, 1993). The farming system is the integration of all agricultural businesses to which a farm family allocates its resources in order to effectively use the already-existing businesses to increase the productivity and profitability of the farm. In IFS, farmers plant crops to ensure efficient and profitable production that is economically viable and produces food that is safe, healthful, and of high quality preserving and improving the environment while growing fresh food and arable crops. In India, the future of tuber crops seems promising. With their ability to thrive in a variety of environments, ability to feed disadvantaged populations with staple foods, ability to produce wholesome animal feed, and ability to produce a variety of farm, home, and industrial products, tuber crops can consistently meet the needs of the public's food and nutritional security.

Tuber Crops based Farming System

After cereals and grain legumes, root and tuber crops are the third most significant food crops for humans. They provide either a staple or supplementary diet for nearly a fifth of the world's population. Due to their high calorific value and carbohydrate content, these crops play a significant role in the food security of developing countries. These crops not only offer a cheap source of energy but also provide other vitamins and minerals. According to reports, these crops yield more dry matter per unit area than any other crop and have the best biological efficiency as food producers. It is amazing how well they can survive during drought circumstances and grow in rainfed environments. Due to the ever-growing population, tuber crops assume considerable relevance in the scope of food and nutrition security. In addition to enhancing human diets, tuber crops have therapeutic characteristics that can treat a wide range of illnesses or reduce their incidence.

Indian agriculture must pursue both horizontal and vertical growth in crop production to meet the country's food needs. Multiple cropping to improve cropping intensity has limited potential because the amount of arable land is decreasing every year, whereas vertical production expansion has enormous potential. Additionally, the majority of crops have reached their genetic yield potential. However, the genetic potential of tuber crops has not yet been fully achieved through sustainable production techniques. The commonly cultivated tropical tuber crops are: Sweet potato (*Ipomoea batatas*), Elephant foot yam (*Amorphophallus paeoniifolius*), Cassava (*Manihot esculenta*), Swamp taro (*Colocasia esculenta var. stoloniferum*), Lesser yam (*Dioscorea esculenta),* Taro (*Colocasia esculenta var. anticorum*), Tannia (*Xanthosoma sagittifolium*), Bunda (*Colocasia esculenta var. esculenta*), Greater yam (*Dioscorea alata*), Aerial yam (*Dioscorea bulbifera*), Yam bean (*Pachyrhizus erosus*), White yam (*Dioscorea rotundata*), Chinese potato (*Solenostemon rotundifolius*), Arrowroot (*Maranta arundinacea*) etc. Tuber crops can use available resources more effectively, especially when there is partial sunlight and delayed moisture (Nedunchezhiyan and Laxminarayana, 2006). These crops also have very adaptable planting and harvesting methods, making them ideal for inclusion in many farming systems. Root and tuber crops-based IFS is a farming system centred on tuber crops comprises the planting of tuber crops alongside other seasonal, horticultural, and silvicultural crops both mixed/ intercropping or sequential cropping; and the use of tubers and leaves in the production of animal growth in both live and processed form. Additional tuber crop by-products, including cassava bagasse, are used in related processes.

To achieve adequate yields and profitability when tuber crops are cultivated in plantations as intercrops, the main crop and the subsidiary crop should be produced using the appropriate management procedures (including manures and fertilisers) as for sole crops.

Tuber Crops based IFS at Andaman & Nicobar Island

According to a study conducted in a tribal area (A&N islands) by the ICAR-Central Island Agricultural Research Institute (CIARI), Port Blair, the major sources of income for tribal people include coconuts, tuber crops, pigs, and marine fishing. Due to a shortage of planting materials and lack of scientific cultivation of tuber crops-based farming, they rarely engage in mixed cropping with coconut, tuber crops, spices, and fruit crops. They solely cultivate Nicobar Aloo (greater yam), mostly for food and to generate up to ₹10,000 in revenue annually. They use to care for their animal and work in their garden for an hour each day. Additionally, they also do fishing for domestic consumption. They have no alternative means of support for their daily needs. The employment days in their traditional system were only 46 man-days/year. After the intervention of the tuber crops-based farming system, the tribal farmers made between ₹5,500 and ₹9,650 from agricultural products, and between ₹20,500 and ₹43,000 from pigs. The system's overall revenue ranges from ₹29,000 to ₹52,650. As a result of the tuber crops-based farming system's intervention they are now working 2 to 3 hours each day, creating 91 to 140 days' worth of employment annually, and earning ₹30,255 more on average for 10 farm families than they were under their previous system. As a result of the successful intervention, more tribal youth are expressing interest in adopting a farming system centred on tuber crops as a source of income.

Tuber Crops based IFS at Chhattisgarh

An investigation on tuber crops-based farming system studies was conducted with different components of farming from 2012-13 to 2015-16 at two locations (Bastar and Kondagaon district) of Bastar division of Chhattisgarh (Shanker et.al., 2018). Of the farming system studies, the socio-economic profiles of the farmers in both areas were documented. The farmers' gross and net revenue improved to ₹5,33,110 and ₹4,03,720 ha^{-1}, respectively, after the intervention of a tuber crops-based integrated farming system in the farmer's field in Gumiyapal village. The integration of fruit, fish, poultry, and ducks along with cereals and pulses led to an increase in income. After the implementation of an integrated agricultural system based on tuber crops, the gross and net income increased by 463% and 610%, respectively. The B:C ratio improved from 1.56 to 4.12. Employment was generated for 261 man-days compared to 199.

The farmers' gross and net income increased to ₹5,28,660 and ₹3,97,960/ha respectively, after the intervention of a tuber crops-based integrated farming system in the farmer's field in Tatirash village. The integrated agricultural method based on tuber crops resulted in 397% and 447% increase in gross and net income, respectively. The B:C ratio improved from 2.18 to 4.04. Different crop, livestock, and fruit crop components were tested on the fields of tribal farmers in the villages of Gumiyapal (Tokapal, Bastar) and Tatirash (Keshkal Kondagaon) as part of farming system studies. When compared to other crops Amorphophallus/Zimikand had the highest B:C ratio (1:3.35), followed by *Dioscorea bulbifera* (1:3.20), and *Colocasia* (Bunda) (1:2.75). The number of man-days employed increased from 203 to 263 man-days. However, in a tuber crops-based integrated farming system, more jobs were created; work was spread out throughout the year. As opposed to monoculture, a farming system based on tuber crops takes an integrated approach. The productivity of the land, farmer income, and the sustainability of the farmer might all grow thanks to this concept. This could serve as a model to draw in and retain the current generation of farmers who are moving to cities. For increased sustainability and production, policies must be designed to help small farmers in extending their activities toward both on- and off-farm pursuits. The model created has the power to change Chhattisgarh's tribal region, particularly the Bastar and Kondagaon districts.

Tuber Crops based IFS at Assam

From 2013-14 to 2015-16, a research programme was carried out in the farmers' fields at Dhenususa, a village in Assam's Jorhat district (Saud, 2016). The farmer's 0.26 ha monoculture of rice field was converted to a tuber crop-based farming method. For increased revenue, employment, and sustainability, several components such as tuber crops, floriculture, fruits, vegetables, fish, duck, and dairy products were integrated. Following the intervention, there was a noticeable increase in the yield of rice equivalent, farm income, and employment. A significant portion of the farm's income came from the tuber crops. Due to its higher productivity and better price, elephant foot yam had the highest rice equivalent yield per unit area among the tuber crops, followed by sweet potato (4.27 kg m^{-2}), greater yam (3.84 kg m^{-2}), and taro (2.77 kg m^{-2}). Sugarcane (chewing type) had the highest productivity (3.55 kg m^{-2}) among the *kharif* field crops in terms of yield comparable to rice. In addition, the banana crop generated the highest rice equivalent yield (8.50 kg m^{-2}) in the IFS system, followed by flowers (5.59 kg m^{-2}), ginger (5.50 kg m^{-2}), winter vegetables (4.66 kg m^{-2}), and lemon (4.25 kg m^{-2}). By transforming the land into a farming system model, the typical rice land's production of 3.33 t ha^{-1}

could be increased to 28.42 t ha^{-1}. This was around 8.5 times greater than the previous approach of mono-cropping rice (861%). The farming system based on tuber crops has the potential to increase employment from the current 212 man-days per hectare in a year to up to 615 days. Therefore, this strategy could transform the monoculture of rice on marginal land holdings into a sustainable farming model that can attract and retain farmers in agriculturally underdeveloped states like Assam.

Tuber Crops based IFS at Tripura

Tuber crops were included in the farming system in the hilly land of Tripura which enhanced the productivity per unit area besides improving food security for small and marginal farmers (Datta *et al.*, 2012). Before the intervention, the 0.48 ha of hilly land (moderate hillock upland) owned by the self-help group Abachanga was left fallow and overgrown with weeds and shrubs. The area was chosen for the production of tuber crops such as Dioscorea, Elephant foot yam, Tapioca, Ginger, Sweet Potato, etc. because it was tilled and without any irrigation supply. Upon clearing, the ground was divided into several small plots for the various tuber crops, each measuring about 800 m^2. Banana suckers were planted around the edges of each allotment. Dioscorea, Elephant foot yam, and Tapioca were planted at 90 cm × 90 cm, sweet potatoes at 60 cm × 20 cm, and ginger at 1.5 tons per hectare. Two piglets of one unit were combined for fattening with tapioca, sweet potatoes, etc. as feed. Farmers provided some concentrate and kitchen garbage for better pig growth. Before the intervention, farmers in the selected locations of the Dhalai District had never heard of a farming system based on tuber crops; there had never been any cultivation of that kind of tuber crop. The farmers lacked knowledge on how to use the fallow land. They would eat the tuber crops during the monsoon season, when there was less food available owing to constant rain, as well as while working on Jhum land after collecting from the jungle. According to their theory, it is a highly powerful and enduring thing. They were making fences out of tapioca (Kathaloo). Additionally, pigs were fed on the tuber crops. Dioscorea, elephant foot yam, tapioca, ginger, and sweet potato all had productive yields of 5.63, 3.75, 1.9, 2.0, and 4.25 tons per hectare, respectively. From 0.48 hectares of land with a B:C ratio of 2.15, farmers may get a net income of ₹20,950. In a single year, a 0.48 ha piece of land produced a net revenue of ₹20,950 from the crop component of ₹14,150 and the pig component of ₹6,200. For the purpose of maintaining the health of the soil, land and various waste products were used effectively.

Conclusion

The productivity of farms can be multiplied many times by switching from cropping to multi-enterprise farming systems. Rainfed regions can only grow crops for four months and frequently experience extremes of bad weather. Farming work is spread out throughout the year due to diversification. The integration of the farm and the house effectively uses family labour, especially female labour. Productivity rises as a result of resource usage that is efficient. Reduced reliance on outside supplies and effective utilisation of family labour all result in lower production costs. Ultimately, lower cultivation costs and higher prices for organically grown goods boost farm profitability. The farming system concept offers smallholder farmers financial security and assurance of a living.

References

Chadha, G.K. (1993). Non-farm employment for the rural households: Evidence and prognosis. Indian Journal of Labour Economics. 38(3): 296-327.

Datta, M., Sankaran, M., Sarkar, R. and Sarma, R. (2012). Tuber Crop-based Farming System for Food Security and Livelihood Improvement in Dhalai, Tripura. In: Site Specific Farming System Options for Rural Livelihood – success stories from NEH Region.

Munda, G.C., Ngachan, S.V., Das, A., Malngiang, S. and Chowdhury, S. (2012). 2012. Site Specific Farming System Options for Rural Livelihood – success stories from NEH Region. NAIP Bulletin No. 2, ICAR Research Complex for NEH Region, Umiam, Meghalaya, pp. 54-55.

Nedunchezhiyan, M. and Laxminarayana, K. (2006). Tuber crops based cropping system in eastern India. Kisan World, November, pp. 45-46.

Saud, B.K., Alam, S., George, J., Hazarika, G.N., Choudhury Talukdar, M., Kotoky, U. and Gogoi, B. (2017). Tuber Crop Based Farming System: An Approach to Boost up Livelihood of Marginal Farmers. Journal of Root Crops. 42(2): 152-157.

Shanker, D., Banjare C. and Sahu, M.K. (2018). Tuber Crops Based Integrated Farming System Studies in Bastar and Kondagaon Districts of Chhattisgarh. Int. J. Curr. Microbiol. App. Sci. 7(9): 1650-1658.

Swaminathan, M.S. (1987). Inaugural address. In: International Symposium on Sustainable Agriculture. International Rice Research Institute, Las Banos, Philippines.

6

Entrepreneurship Development through Fish-Poultry Based Integrated Farming System

Jham Lal, Soibam Khogen Singh and Pradyut Biswas

Department of Aquaculture, College of Fisheries, Central Agricultural University (Imphal), Lembucherra, Tripura

Abstract

Agriculture and livestock agribusiness techniques are assessed as the backbone of the Indian economy. A mixed crop-livestock system known as an integrated farming system comprises at least two distinct but conceptually linked sections of agricultural and livestock businesses. One of the world's greatest agricultural sectors is the chicken business. Poultry-fish farming is an integrated farming technique that involves rearing poultry and fish together, including chicken, duck, as well as geese. Layers are poultry farmed for their eggs, whereas broilers are chicks grown for their meat. Because of the physiological flexibility of the broiler, poultry husbandry is widely practiced as well as frequently associated with chicken throughout the world. As a part of agricultural systems, integrated farming systems consider the ideas of reducing risk, boosting productivity and profits, as well as better utilising biodegradable waste as well as agricultural leftovers. Agricultural leftovers could be utilised for animal nutrition, as well as the production and distribution of animals including their by-products can increase crop production by strengthening the nutrients that promote soil quality whereas lowering the need for synthetic fertilisers. It is necessary to write a strategy for administrators to assess and promote it widely with minimal financial help via short-, medium-, or long-term loan repayments as well as other promotional benefits.

***Keywords*:** Fish, Poultry, Nutrient recycling, IFS and Entrepreneurship development.

Introduction

Agriculture and livestock agribusiness techniques are assessed as the backbone of the Indian economy. Since the livelihood of the population depends on these traditions, the practice is more common in rural regions. Farm owners find it impossible to make a living with traditional farming methods, nevertheless, because of inflation. Integrated farming, a contemporary agricultural practice that increases production and resource effectiveness while lowering the threat to small-scale farming homeowners through crop diversity, is regarded as a sustainable farming approach. Fish farming is an important component of in integrated fish production, which links fish production with agricultural or livestock farming (Ayyappan *et al*., 2011).

One of the world's greatest agricultural sectors is the chicken business. Due to significant changes made to the chicken business over the past few generations, this business has continued to expand (Gerber *et al*., 2007). Poultry-fish farming is an integrated farming technique that involves rearing poultry and fish together, including chicken, duck, as well as geese. The effluent or animal waste produced by poultry is used as food for the fish inside the pond. Farmers benefit from this practice because it reduces costs associated with traditional meals and allows for profit maximisation through feeding fish using poultry feces (Gabriel *et al*., 2007). The International Feed Industry Federation indicates that livestock output will be doubled by 2050. Animals like goats, swine, as well as poultry are the primary sources of meat. It is anticipated that consumption of the latter, which is among the most popular animal meals, will increase by more than 90% by 2050 (Alexandratos and Bruinsma, 2012). Chicken has been identified to become one of the meats with the least impact on the environment, because of the lack of enteric fermentation as well as the reduced feed conversion rate (FCR) (Gerber *et al*., 2007).

The latest advancements in poultry genetics and a comprehensive understanding of their dietary needs have enabled the achievement of extremely high feed utilisation rates (1.5-1.8 g feed/g weight growth). Feeding, feed conversion ratio (FCR), live weight increase, and death rates are the primary determinants of the environmental impact of broilers (Bahadori *et al*., 2017). According to the Food and Agriculture Organisation (FAO), the worldwide chicken population in 2016 was much above 23 billion, and over five times the chicken population 50 years earlier (Mottet and Tempio, 2017). Human population growth, rapid urbanisation and increased purchasing power are the variables that influence increasing demands for chicken products (Mottet and Tempio, 2017). Chicken meat production is expected to increase by over 2.5% annually between 1999-2001 and 2030, and by 1.5% annually between 2030 and 2050, while the demand is projected to increase by around 2.6% and 1.5%, respectively, during the same period (FAO, 2006). Poultry production remains a relatively unexplored sector of farm use practice. Therefore, this agricultural approach is quickly gaining acceptance and popularity among producers. The farm is built on several engineering as well as agricultural production techniques.

Principles of Integrated Farming System

In the integrated farming system, waste is defined as "just a misdirected product that might become a useful property for just another commodity," as well as "there is no waste" (FAO, 1977). A mixed crop-livestock system known as an integrated farming system comprises at least two distinct but conceptually linked sections of agricultural and livestock businesses (Okigbo, 1995). This is a fish farming technique that is combined with poultry as well as uses clean animal manure for feeding fish. It has also been stated the existence of synergistic as well as complementarities among firms that have an agricultural and livestock element that constitute the basis of the Integrated Farming System idea. This theory states that integration typically occurs when the results (mostly metabolic end products) of one business are utilised as input variables by another within the framework of a farming industry (Edwards, 1997; Jitsanguan, 2001). Interconnected farming practice is a combination of animal-crop methods in which the livestock element is frequently grown on agricultural waste materials whereas and the animals are utilised to till the soils and create excrement for use as fertiliser and fuel (Jayanthi *et al*., 2000). As part of agricultural systems, integrated farming systems consider the ideas of reducing risk, boosting productivity and profits, as well as better utilising biodegradable waste as well as agricultural leftovers (Radhamani *et al*., 2003). A type of combined farming method known as an integrated farming system integrates agricultural and livestock businesses in an additional and/or complementary way (Agbonlabor *et al*., 2003). A part of Farming System Research (FSR), integrated farming involves changing farming practices to maximise agricultural output and ensure that commodities are used as efficiently as possible (Jayanthi, 2006). An integrated farming system is a coordinated collection of parts, components, and tasks that producers carry out on their respective properties in accordance with their resources and conditions to sustainably increase productivity and net farm revenue (Singh and Ratan, 2009). In principle, fish-poultry integrated farms usually follow three basic techniques.

Combination of Poultry-Fish Farming

A mix of poultry-fish farming with the cultivation of agricultural items including rice results in more numerous revenue sources; however, this is a complicated farming management approach (Ramanathan *et al.*, 2020). The concept of integrated farming systems includes the incorporation of animal elements. It would increase rural employment production via animal-based foods; minimise pesticide usage by complementary manuring and parasite prevention from livestock elements, and promote dietary diversification, all of which would contribute to improving the nutritional well-being among these resource-poor producers. Strategic integration of any one or several of the operations with such a farming element resulted in greater exploitation of accessible resources by successful reclamation of leftovers or garbage, as well as entrepreneurship for family labour even during the off-season, making their farming a profitable component (Behera *et al.*, 1999).

Cage-Cage Poultry-Fish Farms

Cage-cage poultry and fish farming system, in which the fishes are kept inside a cage in a large water body. With this type of construction, smart farming is much more practical. The difficulty with this method of construction is the risk of fish loss due to netting collapse, particularly in dams that are naturally formed in water bodies (Zaniboni-Filho *et al.*, 2018). To reduce costs and make cage fish farming a sustainable endeavour, a modern production method needs to be developed (Garcia *et al.*, 2013). In addition to ponds, rivers, estuaries, as well as the marine environment, fish enclosures may also be erected there. It is easier to see, feed, as well as capture fish when they are contained in these structures (Beveridge, 2004; Kumar and Karnatak, 2014). Nevertheless, because it is an open platform, interactions in the fish farms have an impact on the surrounding environment and vice versa, thus choosing the site for the enclosures becomes a crucial consideration (Cardia and Lovatelli, 2015).

Table 1: Industrial and Organic Methods of Poultry Production

Criteria	Industrial method	Organic method	References
Maintenance	Free range, natural daylight, as well as clean air are insufficient in cramped chicken houses with cage and flooring confinement	Birds should be given freedom of movement area and also be permitted to move around freely in their natural habitat	Erensoy *et al.* (2016); Roiter *et al.* (2020)
Poultry Density	<25 poultry m^{-2}	<19 poultry m^{-2}	Erensoy *et al.* (2016); Kim *et al.* (2018)

Criteria	Industrial method	Organic method	References
Feeding	The vast majority of feeding bases are composed of compound diets as well as certain formulations, such as those comprising growth hormones and food supplements	The fodder basis contains at least 50% of foodstuff that is produced mostly on farm or that is produced naturally by several other farms within the same area	Rezaei *et al.* (2018); Singh and Bhatt (2021)
Breeding	The most prevalent techniques involve assisted reproduction, genetically modified, and extensive breeding	Natural breeding is assumed, especially among slow-growing bird varieties	Chalova *et al.* (2016); Roiter *et al.* (2020)
Health Status	To control and treat chicken infections, antibiotics, as well as other potent medicines, are utilised	Immuno-biological medications are permitted for use in disease prevention, as well as the treatment of chickens, organically based phytotherapeutic, homeopathic formulations, and microelements are used	Dhama *et al.* (2014); Souillard *et al.* (2019)
Economic Activity	Profit maximisation is achieved by increasing production and cutting down on the time it takes to produce goods that can be sold	Utilising a pricing structure, maximizing profits to greater, sustainable products	Roiter *et al.* (2020)
Separate Space	Not separate space for infestation poultry	Separate space for infestation poultry	Kim *et al.* (2018)

Beveridge (2004) states the following as three important factors to take into account while selecting a position:

- Environmental standards for fish production include topographic features, hydrology, dissolved oxygen concentration, temperature, alkalinity, visibility, as well as algae growth.
- Climate (precipitation and temperature changes), current flow, depths, as well as substrates are environmental factors for fish cage designs.
- Regulations, preparation, accessibility, access to the marketplace, as well as safety are all examples of logistics and management factors (Zaniboni-Filho *et al.*, 2018).

Fish cum Poultry Farming

Pond poultry production involves placing bird confinements inside or near the pond, and excrement is delivered to the freshwater ecosystem units (either manually collected or through chicks moving inside the water body) (Gabriel *et al.*, 2007; Ramanathan *et al.*, 2020). Poultry-fish farming combines raising fish alongside poultry livestock including chickens, ducks, and geese (Tokrisna, 1995.) Fish and chicken farming are popular activities due to their sales and profits. Fish farming can include the integration of birds reared for their eggs (layers) or for meat (broilers). This one will lower the cost of inputs, including feed and fertiliser, to increase profitability (Asala, 1994). If necessary, the chicken residence could be built within or elevated above the pond. Bird excretions can either be eaten by fish directly or used as manure to nourish the pond. In order to maximise the utilisation of the land, it would be more suitable as well as superior if the chicken residence were elevated over the ponds. This would permit vertical integration as well as the transportation facilities of faeces to waterbodies. Excrement will fall directly into the ponds, making the chicken coop environmentally friendly (Gabriel *et al.*, 2007).

Fig. 1: Fish-Poultry Integrated Farming System

Integrated farming systems involve the combination of various agricultural components to create a synergistic and sustainable system. When it comes to poultry-fish integration, it refers to the integration of poultry farming with fish cultivation. This integrated system allows for the efficient use of resources, including the utilisation of poultry waste as a nutrient source for fish production, while also providing additional income streams for farmers.

1. **Nutrient Recycling:** Poultry waste, such as manure and uneaten feed, can be utilised as a nutrient source for fish ponds. This recycling of nutrients reduces waste and the need for external inputs, such as commercial fish feed or chemical fertilisers.

2. **Increased Productivity:** The combination of poultry and fish farming can enhance overall productivity. Poultry can provide a source of protein-rich feed for the fish, leading to improved fish growth rates and increased fish production.
3. **Diversification of Income:** Integrating poultry and fish farming allows farmers to diversify their income sources. They can generate revenue from both poultry products (e.g., eggs, meat) and fish sales, reducing the risks associated with relying solely on one enterprise.
4. **Efficient Resource Utilisation:** Integrated systems make efficient use of land, water, and other resources. For example, the water used in poultry farming can be recycled and utilised in fish ponds, minimising water wastage.
5. **Environmental Sustainability:** The integration of poultry and fish farming can contribute to environmental sustainability by reducing the environmental impact of waste disposal and optimising resource utilisation.

Conceptualising Poultry Farming

The practice of raising domesticated birds for their eggs and meat is known as poultry farming. The most popular poultry raised is chicken because it is most widely consumed, generates greater revenue, and requires minimal upkeep. Layers are poultry farmed for their eggs, while broilers are chicks raised for their meat. Due to the physiological adaptability of broilers, poultry husbandry is widely practiced and often associated with chicken throughout the world. Such cultivation is enhanced by the opportunity to start the business with little initial investment. Poultry, which includes birds such as chickens, ducks, turkeys, geese, and quails as well as other animals like pigeons raised for food or wings, derives its name from the French/Norman term "poule," which comes from of the Latin word "pullus," meaning little creature. Poultry refers to the rearing of birds for both meat and eggs as opposed to hunting or sports purposes. Guinea pigs are even used in some nations' poultry farming, which involves domesticating as well as keeping birds in cages for use in commerce and human consumption. Poultry production is a common activity in both urban and rural locations and has a significant positive impact on economic development. People raise chickens all around the world for a living as well as to involve the women in the household throughout the growing of the birds. Cultivating cultured birds for eggs and meat is referred to as poultry. Backyard poultry raising is prevalent as well as acknowledged as a profession in several rural places (Mitra *et al.*, 2021).

Table 2: Plant and Animal Protein Source Used in Natural Poultry Farming

Protein source	Examples	Characteristics
Grain legumes	Soybean meal, Faba beans, Lupines, Grass peas	Rich in lysine; unsuitable as a sole source of animal protein; anti-nutrients present
Cereal grains	Corn, Corn hybrid	High amount of methionine and crude protein
Fish	Fish meal	Methionine abundance and an appropriate amino acid profile
Oilseeds	Sunflower meal, Canola meal, Sesame meal, Soybean oil	Comparatively high protein percentage; a great source of the majority of the essential amino acids; rich sulphur-containing amino acid content; greater leftover fat; the existence of anti-nutrients
Microorganisms	Bacteria, Yeasts	Improved methionine output
Insects	Tree ants, House crickets, Black soldier fly, Common housefly, yellow mealworm, grasshoppers, locusts, crickets, silkworms	Highly varied nutritional content; protein, lipid, vitamin, as well as mineral content; potential transporters of toxic substances, poisons, contaminants, and/or pathogens

Source: Yadav *et al.* (2022)

Recycling of Waste and Resource Utilisation

People often equate it with a mixture of farming methods, but there are too many differences between this and an integrated farming system. The differentiation between the two is that an integrated farming system's businesses depend on one another and provide assistance for one another. A combined farming practice includes elements like agriculture and livestock which subsist separately from one another. In this farming, combining crops as well as cattle generally reduces uncertainty rather than recycling materials (Csavas, 1992). While in interconnected farming methods, animals and crops work together to generate a synergistic, as well as recycle enabling the best possible utilisation of the commodities. Agricultural leftovers could be utilised for animal nutrition, as well as the production and distribution of animals including their by-products can increase crop production by strengthening the nutrients that promote soil quality whereas lowering the need for synthetic fertilisers. Although a high crop-livestock combination is frequently viewed as a positive development, small farmers must have adequate access to information, resources, as well as supplies to maintain this system in a manner that is long-term both environmentally and economically stable (FAO, 1977). The level of cooperation and collaboration inside the crop production system determines how interconnected farming practices differ from a commercial farming method rather than being an ultimate differentiation (Tipraqsa, 2006).

The integration of waste recycling and resource utilisation in a fish-poultry integrated system can provide numerous benefits, including efficient use of resources, waste management, and increased sustainability.

1. **Waste Recycling:** The system can utilise various waste streams as inputs for different components, minimising waste and promoting circularity.
   a) **Fish Waste:** Fish waste, such as fish offal or uneaten fish feed, can be converted into valuable resources. For example, it can be processed into fishmeal, which can serve as a protein-rich feed ingredient for poultry.
   b) **Poultry Waste:** Poultry manure contains nutrients that can be utilised as fertilisers or converted into biogas through anaerobic digestion. Biogas can be used as an energy source for heating or electricity generation.
2. **Resource Utilisation:** The integrated system can optimize the use of resources by creating symbiotic relationships between fish and poultry.
   a) **Nutrient Cycling:** Fish produce waste in the form of ammonia, which can be converted into nitrate by nitrifying bacteria. Nitrate-rich water can then be used to fertilize plants or hydroponic systems, which can serve as a feed source for poultry.
   b) **Aquaponics:** Aquaponics combines fish farming (aquaculture) with hydroponics, where plants are grown in water without soil. The nutrient-rich fish waste provides fertilization for the plants, while the plants help filter and purify the water for the fish.
3. **Water Management:** By integrating fish and poultry systems, water usage can be optimized. For example, water from the fish tank can be circulated through the poultry area for cleaning and then returned to the fish tank. This minimises water wastage and ensures efficient water use.
4. **Environmental Benefits:** The integrated system can help reduce the environmental impact associated with waste disposal. By recycling waste and utilising resources efficiently, it decreases the amount of waste that needs to be treated or disposed of outside the system.
5. **Economic Viability:** The integration of fish and poultry systems can enhance the economic viability of the overall operation. The utilisation of waste as valuable resources reduces the need for external inputs, thereby reducing costs. Additionally, the sale of fish and poultry products can generate additional income streams.

Table 3: Nutrient Recycling Fish-Poultry Integrated Farming System in India

Farming System	Nutrient Reclaiming (kg)			Reference
	N	P_2O_5	K_2O	
Crop + Fish + Poultry	192.5	119.7	77.8	Sanjeev *et al.* (2011)
Crops + Horticulture + Cattle + Fishery + Poultry + Apiary	359.4	140.0	398.6	Ravisankar *et al.* (2016)
Crop + Horticulture + Dairy + Sheep + Poultry	91.0	42.0	75.0	Goverdhan *et al.* (2020)
Crop + Dairy + Fish + Poultry	55.0	17.0	76.0	Paramesh *et al.* (2021)

Prospects of an Integrated Farming System

Integrated fish farming techniques save money since leftovers from one farming method are used to feed another's nutritional needs. Fish farming practices can be combined with agri-aquaculture-based processes (paddy-based fish farming, horticulture-based fish farming, mushroom-based fish farming, seri-based fish farming, and vermicompost-based fish farming), as well as livestock-fish-based farming techniques (cattle-based fish farming, pig-based fish farming, goat- or sheep-based fish farming, poultry-based fish farming, and duck-based fish farming) (Ayyappan *et al.*, 2011). A multidisciplinary strategy for addressing the issues of food shortages is the integrated farming system. By increasing food production, and revenue, as well as the resiliency of the soil's reproductive ability, integrated farming systems, enable smaller farms and farmers with limited resources to combat issues of food insecurity and malnutrition. The viability and profitability of farming methods are seriously threatened by the trend of shrinking land availability. Under these circumstances, integrating land-based businesses like agribusiness, horticulture, chicken, ducks, piggeries, as well as fisheries within the farm is suitable (Gurjar and Swami, 2019).

The integration of fish and poultry farming has gained attention as an efficient and sustainable agricultural system. Here are some prospects and potential benefits of fish-poultry integrated farming –

1. **Efficient Nutrient Utilisation:** Fish and poultry have different dietary requirements, and their waste products can be effectively utilised in an integrated system. The nutrient-rich wastewater from fish tanks can be used as a nutrient source for the production of feed and fertilisers for poultry, reducing the need for external inputs (Yun *et al.*, 2016).
2. **Increased Income and Diversification:** Integration of fish and poultry farming allows farmers to diversify their income streams. They can generate revenue from both fish and poultry products, such as fish,

eggs, meat, and feathers. This diversification can contribute to financial stability and risk reduction (Brum *et al.*, 2016).

3. **Improved Resource Utilisation:** The combination of fish and poultry farming optimizes the use of resources such as land and water. Poultry houses can be built over fish ponds, utilising the space more efficiently. The droppings from poultry serve as a nutrient source for fish, reducing the need for artificial fertilisers and enhancing water quality (Saha *et al.*, 2019).
4. **Disease and Pest Control:** In an integrated system, the presence of fish can help control pests and parasites that affect poultry. Fish feed on mosquito larvae and other insects, reducing the pest population around the poultry farm. This natural pest control approach reduces the reliance on chemical pesticides (Alam and Khatun, 2015).
5. **Environmental Sustainability:** Fish-poultry integrated farming can contribute to environmental sustainability. The nutrient cycling between fish and poultry reduces nutrient pollution in water bodies, and the efficient use of resources minimises the ecological footprint of the farming system (Verma *et al.*, 2017).

Table 4: Economic Analysis of Fish-Poultry Integrated Fish Culture (Farmers' Practice) from an average 0.2 ha Area of Pond

Component	Item	Poultry cum Fish Farming		
		Quantity	Rate (₹)	Cost
Fish	Pond preparation	-	-	1,000.00
	Lime	50 kg	7/kg	350.00
	Cow dung	20 q	25/q	500.00
	Seed	5000 no	250/1000	1,250.00
	Feed	-	-	-
	Labour	-	-	3,000.00
	Miscellaneous	-	-	4,000.00
Poultry	Chicks	50 no	16/chick	800.00
	Medicines	-	-	200.00
	Feed	8qtl	1800	14,400.00
	Miscellaneous	-	-	300.00
A. Total expenditure				**25,800.00**
Income from fish		1288.28 kg	85/kg	1,09,503.00
Income from egg		2250 egg	5/egg	11,250.00
Income from poultry meat		46 bird	300	13,800.00
B. Total income				**1,47,553.00**
Net income (B-A)				**1,21,753.00**

Source: Sharma *et al.* (2016)

Importance of an Integrated Farming System

In order to minimise the negative impacts of intensive farming while protecting the natural environment, an integrated farming system provides resource-saving techniques (Gupta *et al.*, 2012). To help nutritionally challenged individuals get greater animal protein as well as other components, tiny integrated fish farming practices can be employed as a helpful strategy (Elvevoll and James, 2000). This farming method boosts agricultural production as well as overall household incomes while also enhancing nutrition, promoting the use of environmental assets, and ensuring resource sustainability. Given that it minimises expenditures for fertilisers as well as feeding in the fish culture ponds and maximises advantages, such a combination of chicken and fish farming may be a financially successful and profitable method. The practice of directly using fresh animal wastes in the aquacultural pond is known as the combined aquaculture industry, and it is an environmentally responsible way to meet the different needs of farm families while conserving the natural resource and maintaining good sustainability of the environment (Little and Edwards, 1999). Homestead ponds can be yearly or perennial, but they are particularly prevalent in communities. In rural India, keeping chickens in the backyard is a long-standing custom that can be combined with raising fish in household ponds. Poultry dung is a full-spectrum fertiliser that has elements of both organic and inorganic fertilisers. Because chickens have very limited digestive systems, 80% of their feces may be included as non-digestible feed materials, which may also comprise 20–30% of their protein content (Chen, 1981; Pudadera Jr. *et al.*, 1986). One of the leading ways to promote rural women is the integrated fish farming strategy, which makes use of home resource management.

Synergistic Effects of Fish-Poultry Integration

In an integrated aquaculture-agriculture (IAA) model, improved water accessibility year-round can also enhance agricultural production (Ahmed *et al.* 2014). For most IAA processes, the primary pond inputs—particularly feed and/or fertilisers come from agricultural waste (Brummett and Jamu 2011; Ahmed *et al.* 2014). The fish ponds may play a crucial part in promoting various operations in these kinds of ecosystems. For instance, during the dry period, nutrient rich pond water can be used to irrigate vegetables, while pond sediment can be utilised to fertilise nutrient-depleted farmland to boost the yield of already-planted crops (Ahmed *et al.* 2014; Limbu *et al.* 2017). In *Oreochromis niloticus*-livestock interconnection, including *O. niloticus* poultry production, chicken excrement high in nitrogen, phosphorus, as well as potassium serves as fertilisers to promote the growth of naturally occurring aquatic food organisms (Little and Edwards 2003). Fish's biological sources of food include primary production, zooplankton, as well as bacteria; however, in fish-livestock assimilation, O. niloticus may consume chicken excrement directly. There is no need for supplemental feeding during this method of O. niloticus cultivation (Mridha *et al.* 2017). Farming depends mostly on synergistic interactions where one thing becomes another's inputs (often by-products or waste) (Brummett and Jamu, 2011). O. niloticus grows more quickly under such an interconnected farming method, increasing output as well as revenue (Limbu *et al*., 2017). Additionally, it creates jobs from a variety of products because of their interaction and synergy. Although waste materials are recycled, the integration of agribusiness further promotes sustainability of the environment by reducing the likelihood of pollutants having negative environmental effects such as contamination (Jayanthi *et al.* 2003; Ahmed *et al.* 2014).

Challenges and Opportunities in Integrated Farming Systems

Integrated farming systems (IFS) offer a holistic approach to agricultural production by combining different farming components to maximise resource utilisation, increase productivity, and enhance sustainability. This system integrates various components such as crops, livestock, aquaculture, and agroforestry in a synergistic manner. Here are some challenges and opportunities associated with integrated farming systems (Nair, 2012).

A. Challenges

1. **Knowledge and Skills:** Implementing an integrated farming system requires farmers to possess diverse knowledge and skills across multiple domains, including crop production, animal husbandry, aquaculture management, and agroforestry. Lack of expertise and training can pose a significant challenge.
2. **Management Complexity:** Managing multiple farming components and their interactions can be complex, requiring careful planning and coordination. Farmers need to understand the specific requirements, interactions, and potential conflicts between different components to optimize resource allocation and ensure the system's success.
3. **Infrastructure and Space:** Integrated farming systems often require adequate infrastructure and space to accommodate different components. Farmers may face challenges in terms of land availability, access to water resources, and the establishment of appropriate infrastructure for livestock, aquaculture ponds, and agroforestry plantations.

4. **Market Access and Value Chain Integration:** Integrated farming systems may produce diverse products, including crops, livestock, fish, and timber. However, accessing appropriate markets and establishing efficient value chains for multiple products can be challenging. Farmers need to develop market linkages and ensure adequate demand and pricing for their various products.

B. Opportunities

1. **Increased Resource Efficiency:** Integrated farming systems capitalize on the synergistic interactions between different components, leading to efficient utilisation of resources such as land, water, and nutrients. For example, animal waste can be recycled as organic fertiliser for crop production, and crop residues can be used as feed for livestock, thereby reducing waste and enhancing resource efficiency.
2. **Risk Diversification:** By diversifying their farming enterprises, farmers can reduce the risks associated with relying solely on a single crop or livestock species. Integrated farming systems provide multiple income streams and can buffer farmers against market fluctuations, climate variability, and disease outbreaks.
3. **Nutrient Cycling and Soil Health:** The integration of crops, livestock and agroforestry in an integrated farming system facilitates nutrient cycling and enhances soil health. Animal manure and crop residues contribute to organic matter and improve soil fertility, while agroforestry systems promote soil conservation and biodiversity.
4. **Climate Change Adaptation and Mitigation:** Integrated farming systems have the potential to enhance climate change resilience and contribute to mitigation efforts. Agroforestry components sequester carbon dioxide, while diverse cropping systems can be more resilient to extreme weather events such as droughts or floods (ICAR, 2019; Ram and Das, 2020; Sinha and Dey, 2021).

Constraints in Integrated Farming System

Implementing an integrated farming system can face certain constraints that may vary depending on the specific context and region. Here are some commonly encountered constraints in integrated farming systems, along with relevant references for further reading:

1. **Limited Access to Capital and Credit:** Lack of financial resources and access to credit can hinder the adoption and expansion of integrated

farming systems. Farmers may face challenges in acquiring the necessary funds for infrastructure development, livestock purchase, and initial investment (Banik and Midmore, 2015).

2. **Knowledge and Technical Know-How:** Successful implementation of integrated farming systems requires specialised knowledge and technical expertise in various agricultural practices. Farmers may lack the necessary training and awareness about integrated farming techniques, such as agroforestry, crop-livestock interactions, and sustainable resource management (Alemu *et al.*, 2021).
3. **Lack of Appropriate Policies and Institutional Support:** The absence of supportive policies, regulations, and institutional frameworks can act as barriers to the adoption of integrated farming systems. Limited government support, inadequate extension services, and ambiguous land tenure policies can hinder the scaling up and mainstreaming of integrated farming approaches (van Noordwijk *et al.*, 2018).
4. **Market Challenges and Value Chain Integration:** Integrated farming systems often produce a diverse range of products, which can pose challenges in marketing and value chain integration. Farmers may struggle to identify appropriate market channels, establish linkages with buyers, and ensure fair prices for their products (Kiringe and Njuki, 2016).
5. **Limited Access to Inputs and Technology:** Access to quality seeds fertilisers, veterinary services, and other agricultural inputs can be a constraint for farmers practicing integrated farming. Additionally the availability of appropriate and affordable technologies tailored to integrated farming systems may be limited (Place *et al.*, 2019).

The significance of the Integrated Farming System was noted to be constrained in the ways listed below. IFS's primary financial restriction is a low budget (Banerjee *et al*, 1990). The primary production restrictions in IFS are the year-round shortage of livestock feed and the absence of labour during periods of necessity (Ngambeki *et al.*, 1992). Resource-poor farmers are constrained from making larger initial investments because they need rapid financial rewards to finance family basic needs such as food, schools, medical services, and loan amounts (Thamrongwarangkul, 2001). The higher initial expenses may prevent producers from converting to integrated farming as well as from taking use of resources integration's advantages (Tipraqsa *et al.*, 2007). Difficulties in obtaining better cattle breeds, regular procurement of fish seed as well as feeds, inexpensive energy-efficient pumped equipment, knowledge of government initiatives, and financial assistance from financial organisations (Nageswaran *et al.*, 2009). IFS's limitations include the high cost of concentrated diet and the scarcity of green manure (40%) while 30% of participants indicated a lack of market infrastructure and cooperative societies. In IFS, 20%, 6%, and 4% of participants, respectively, identified a lack of scientific expertise in livestock farming, scarcity of superior species in the market, and absence of financial assistance (Kadam *et al.*, 2010). As a result, it is necessary to develop a data source on cultivation systems in connection with the many types of farming systems, infrastructures, finance, sustainability, etc. in various cultivation contexts. It is necessary to establish investigation components on agricultural systems with various tenure sizes as well as both financially and socially feasible techniques. Evaluating and improving various technologies created at the research facility mostly in the cultivators' sector. In order to address meteorological whims and climate hazards in various farming contexts, contingency planning is required. It is necessary to write a strategy for administrators to assess in order to promote it widely with minimal financial help via short-, medium-, or long-term loan repayments as well as other promotional benefits (Pandey *et al.*, 2018).

Case Studies of Fish-Poultry Integrated Farming

Begum *et al.* (2019) study shows that women's engagement in entrepreneurship is critical for a nation's economic success, and their involvement in income generating activities increases their level of independence. Women entrepreneurs are considered one of the most important driving forces for Bangladesh's overall growth and long-term income progress. Women's participation in various entrepreneurial ventures, especially in agriculture, has not only strengthened them physically and socially but also financially. The poultry industry including chicken rearing and poultry shelter maintenance in Bangladesh, has entered a new phase, presenting significant potential for the nation. The increasing number of poultry farms and significant increase in overall turnover have encouraged women and businessmen to invest more heavily in this industry. As a result, the over entrepreneurial input has contributed to the country's GDP (Gross Domestic Product). Female entrepreneurship in the poultry sector has evolved through various stages. Despite some notable progress, women continue to face numerous challenges in economic, social, commercial, and technological aspects that need to be addressed promptly. Additionally, appropriate training and technical assistance are necessary to develop a skilled female work force. Most importantly, financial support at both national and international levels is crucial for promoting female empowerment and ensuring sustainable rural development in Bangladesh (Begum *et al.* 2019).

Morrison *et al.* (2006) reported on the management of a contractual chicken farming system in Sarawak, Malaysia, within a state-subsidised context that includes a small residential commodities submarket. The contract's primary goal is to boost earnings as well as educate general entrepreneurial skills for 'bumiputra' minorities as part of an employment equity strategy in remediating the equilibrium in rural improvement and cultural diversity in Malaysia. Few of these programmes have been created specifically to encourage entrepreneurship, although there is a strong interest in encouraging entrepreneurship in indigenous communities at both a conceptual and practical level. Indigenous farm owners make up the majority of small farmers in several emerging economies which are progressively controlled by commodity contracts. In the short term, the poultry production agreement managed by the Sarawak Economic Development Corporation (SEDC) has been evaluated by its capacity to increase the earnings of low-income rural agricultural households rather than by its capacity to turn bumiputra minority, smallholder farmers, into fully formed commercial entrepreneurs. More importantly, the privatized chicken industry is already congested and mostly dominated by vertically integrated, ethnic Chinese-owned businesses that operate in a state-regulated marketplace. They are shielded from the competitiveness of larger farmers in the Peninsular and around the world. There are strong economic interests in keeping the protective measures in place (Morrison *et al.* 2006).

The majority of women in rural areas, especially in the poorer sectors of society, are uneducated and oppressed; therefore, efforts to empower women must focus on reducing oppression, promoting gender awareness, expanding their horizons, and advancing microenterprises. The concept of women's empowerment has many facets and dimensions. It is a process through which women are given more power over household decisions and greater access to financial resources. According to reports, the family's children benefit when the female member of the household earns money (Shettar and Rajeshwari, 2015). Women's empowerment through participation in Self-Help Groups (SHGs) would benefit not just the individual woman but also their families, which would ultimately lead to the growth of society (Thangamani and Muthuselvi, 2013). There is considerable potential for improving rural women's socioeconomic situation through the implementation of well-founded user-friendly technology that promotes self-employment as well as economic growth (Paul *et al.* 2018).

Odinwa and Amugo (2022) study the women's participation and role in tackling agricultural difficulties, which are quite dominant as well as conspicuous. As a result, their importance as well as usefulness cannot be overstated. They are the driving force behind Africa's agricultural production, thus any field of cultivation which is devoid of women because of gender inequalities loses (Odinwa *et al*., 2016). These gender inequalities in agricultural engagement may emerge as a result of conventions, ideas, and mind-sets that confine women primarily to domestic domains, as well as laws that restrict women's ability to obtain land, finance, and factors of production, profession, learning, or medical treatment. Involvement in any activity, such as chicken farming, is contingent on the cost associated and the anticipated rewards, as well as enough publicity to develop an all-around business, granted the advancement adheres to the customs and traditions of the people, particularly women (Odinwa *et al*., 2016). According to studies, rural women contribute significantly to agricultural cultivation in Nigeria as a whole, as well as in Rivers State in particular. The state is primarily populated by a significant number of farmers, the majority of whom are women, and it has a large amount of land appropriate for agriculture and livestock production, especially chicken and fish farming. Nonetheless, about 75% of the internal consumption of poultry meat in Rivers State comes from neighbouring states, with the remainder imported (Odinwa and Amugo, 2022).

Conclusion

Unemployment increases day by day due to the increase in the world's population. Therefore, adopting Integrated Farming Systems can generate entrepreneurship and profitability all over the world. IFS should always be taken into consideration as a well-known option for smallholder farmers. The use of poultry-fish farming in domestic ponds is a commercially viable and practicable technique that might successfully address the challenges of sustainable practices, women's empowerment, as well as livelihood improvement. Integrated fish farming in domestic ponds with enhanced chicken varieties would assist the empowerment of rural women. Agriculture and more precisely chicken agribusiness will reasonably exist as one of their leading interests to them. It has been proven to be more economical than a unilateral cultivation system, helping to spread the economic burden, reduce waste and promote better environmental sustainability.

References

Agbonlahor, M.U., Aromolaran, A.B. and Aiboni, V.I. (2003). Sustainable soil management practices in small farms of Southern Nigeria: A poultry-food crop integrated farming approach. Journal of Sustainable Agriculture. 22(4): 51-62.

Ahmed, N., Ward, J.D. and Saint, C.P. (2014). Can integrated aquaculture-agriculture (IAA) produce "more crop per drop"? Food security. 6(6): 767-779.

Alam, M.J. and Khatun, M.A. (2015). Integrating fish and poultry farming system: A means of improving productivity and profitability. International Journal of Biosciences. 6(11): 235-241.

Alemu, M.T., Vranken, L., Minten, B. and Aynekulu, E. (2021). Tree-based integrated agriculture and agroforestry: An empirical exploration of the enabling environment in Ethiopia. Land Use Policy. 100: 104923.

Alexandratos, N. and Bruinsma, J. (2012). World agriculture towards 2030/2050: the 2012 revision, ESA Working Paper No. 12-03, FAO, Rome.

Asala, G. (1994). Principles of integrated aquaculture. Olatunde AA. In Lake Kaiinji. In: Ayeni JSO, editor. Helmited Guinea Fowl (NMGP).

Ayyappan, S., Moza, U., Gopalakrishan, A., Meenakumari, B., Jena, J.K. and Pandey, A.K. (2011). Handbook of Fisheries and Aquaculture. Edn 2nd, Directorate of knowledge management in agriculture, Indian Council of Agriculture Research, New Delhi, pp. 439-440.

Bahadori, Z., Esmaielzadeh, L., Karimi-Torshizi, M.A., Seidavi, A., Olivares, J., Rojas, S. and López, S. (2017). The effect of earthworm (Eisenia foetida) meal with vermi-humus on growth performance, hematology, immunity, intestinal microbiota, carcass characteristics, and meat quality of broiler chickens. Livestock Science. 202: 74-81.

Banerjee, B.N., Sarker, S.C. and Maity, A.K. (1990). Impact of resource optimisation on cropping pattern and income on 'crop-dairy mixed farm'. Indian Journal of Dairy Science. 43(3): 295-301.

Banik, P. and Midmore, D.J. (2015). Constraints and opportunities for diversification of smallholder mixed farming systems: A case study from Nepal. Agroecology and Sustainable Food Systems. 39(3): 294-318.

Begum, M., Sultana, T. and Islam, I.B. (2019). A Critical Overview on Women Entrepreneurship through Poultry farming in Bangladesh. Journal of Emerging Technologies and Innovative Research. 6(6): 888-896.

Behera, U.K. and Mahapatra, I.C. (1999). Income and employment generation for small and marginal farmers through integrated farming systems. Indian Journal of Agronomy. 44(3): 431-439.

Bell, L.W. and Moore, A.D. (2012). Integrated crop–livestock systems in Australian agriculture: Trends, drivers and implications. Agricultural Systems. 111: 1-12.

Beveridge, M.C.M. (2004). Cage aquaculture. 3rd ed. Oxford: Blackwell Publishing Ltd., Garsington Road, Oxford, UK. http://dx.doi.org/10.1002/9780470995761.

Brum, M., Pereira, R.T., Tavares-Dias, M. and Sousa, G.V. (2016). Integrated fish-poultry production systems: A review of internationally available information. Revista Brasileira de Zootecnia. 45(8): 499-510.

Brummett, R.E. and Jamu, D.M. (2011). From researcher to farmer: Partnerships in integrated aquaculture – Agriculture systems in Malawi and Cameroon. International Journal of Agricultural Sustainability. 9(1): 282-289.

Cardia, F. and Lovatelli, A. (2015). Aquaculture operations in floating HDPE cages: a field handbook. Rome: FAO, FAO Fisheries and Aquaculture Technical Paper, vol. 593.

Chalova, V.I., Kim, J., Patterson, P.H., Ricke, S.C. and Kim, W.K. (2016). Reduction of nitrogen excretion and emission in poultry: A review for organic poultry. Journal of Environmental Science and Health (Part B). 51(4): 230-235.

Chen, Y. (1981). Chicken farming in integrated fish farming vol. 11 pp. 4-30. Regional Aquaculture Lead Center Wuxi, China.

Csavas, I. (1992). Regional review on livestock-fish production systems in Asia. In: Mukherjee, T.K., Moi, P.S., Panandam, J.M., and Yang, Y.S. (Eds.), Proceedings of the FAO/IPT Workshop.

Edwards, P. (1997). Sustainable food production through aquaculture. Aquaculture Asia. 2(1): 4-7.

Elvevoll, E.O. and James, D.G. (2000). Potential benefits of fish for maternal, foetal and neonatal nutrition: a review of the literature. Food, nutrition and agriculture. 27: 28-39.

Erensoy, K., Bayraktar, H., and Tuğalay, Ç.Ş. (2016). The impacts of national and international standards on organic broiler production in developing countries: Turkey case [Conference poster]. In: 3rd International Poultry Meat Congress, 22-26 April 2015, Antalya, Turkey. Proceedings (pp. 220-225). Beyaz Et Sanayicilerive Damızlıkçıları Birligi Dernegi (BESD-BIR).

FAO, China. (1977). Recycling of organic wastes in agriculture. FAO Soil Bull., 40 - Rome. 13. Food and Agriculture Organisation of the United Nations. 2001. Mixed Crop-Livestock Farming: A Review of Traditional Technologies based on Literature and Field Experience. Animal Production and Health Papers, 152. Rome.

FAO, Rome. (2006). Prospects for food, nutrition, agriculture and major commodity groups. World Agriculture: Towards 2030, 2050.

Gabriel, U.U., Akinrotimi, O.A., Bekibele, D.O., Anyanwu, P.E. and Onunkwo, D.N. (2007). Economic benefit and ecological efficiency of integrated fish farming in Nigeria. Scientific Research and Essay. 2(8): 302-308.

Garcia, F., Romera, D.M., Gozi, K.S., Onaka, E.M., Fonseca, F.S., Schalch, S.H. and Portella, M.C. (2013). Stocking density of Nile tilapia in cages placed in a hydroelectric reservoir. Aquaculture. 410: 51-56.

Gerber, P., Opio, C. and Steinfeld, H. (2007). Poultry production and the environment – A review. FAO, pp. 1-27.

Goverdhan, M., Kumari, C.P., Reddy, G.K., Sridevi, S., Alibaba, M.D., Chiranjeevi, K. and Kumar, M.S. (2020). Evaluation of integrated farming system model for resource recycling and livelihood security of small and marginal farmers of Telangana State, India. Current Journal of Applied Science and Technology. 39(34): 17-26.

Gupta, V., Rai, P.K. and Risam, K.S. (2012). Integrated crop-livestock farming systems: A strategy for resource conservation and environmental sustainability. Indian Research Journal of Extension Education. 2(Special Issue): 49-54.

Gurjar, G.N. and Swami, S. (2019). Integrated farming systems of valley regions: Food and nutritional security. International Journal of Chemical Studies. 7(2): 773-778.

ICAR (2019). "Integrated Farming Systems for Sustainable Agriculture", Indian Council of Agricultural Research, New Delhi.

Jayanthi, C., Mythili, S., Balusamy, M., Sakthivel, N. and Sankaran, N. (2003). Integrated nutrient management through residue recycling in lowland integrated farming systems. Madras Agricultural Journal. 90(1/3): 103-107.

Jayanthi, C., Rangasamy, A. and Chinnusamy, C. (2000). Water budgeting for components in lowland integrated farming systems. Madras Agricultural Journal. 87(7/9): 411-414.

Jayanthi, C., Sakthivel, N., Sankaran, N. and Thiyagarajan, T.M. (2006). Integrated farming system: A path to sustainable agriculture. Tamil Nadu Agricultural Univ., Coimbatore, India.

Jitsanguan, T. (2001). Sustainable agricultural systems for small-scale farmers in Thailand: implications for the environment (pp. 1-11). Kawana, Japan: Food and Fertiliser Technology Center.

Kadam, S.S., Hatey, A.A., Nikam, T.R., Landge, S.P. and Palampalley, H.Y. (2010). Constraints of IFS in Kankan region of Maharashtra – A case study. In: 22nd National Seminar on "Role of Extension in Integrated Farming Systems for Sustainable Rural Livelihood, 9th to 10th Dec, 2010, Maharashtra (Vol. 101).

Kim, Y.J., Park, J.H. and Seo, K.H. (2018). Comparison of the loads and antibiotic-resistance profiles of Enterococcus species from conventional and organic chicken carcasses in South Korea. Poultry Science. 97(1): 271-278.

Kiringe, J.W. and Njuki, J.M. (2016). Agroforestry and value chain development in Kenya: The case of fodder production, marketing, and utilisation. Agroforestry Systems. 90(2): 307-321.

Kumar, V. and Karnatak, G. (2014). Engineering consideration for cage aquaculture. IOSR Journal of Engineering. 4(6): 11-18.

Limbu, S.M., Shoko, A.P., Lamtane, H.A., Kishe-Machumu, M.A., Joram, M.C., Mbonde, A. S. and Mgaya, Y.D. (2017). Fish polyculture system integrated with vegetable farming improves yield and economic benefits of small-scale farmers. Aquaculture Research. 48(7): 3631-3644.

Little, D. and Edwards, P. (2003). Integrated livestock-fish farming systems. Food and Agriculture Organisation, Rome.

Little, D.C. and Edwards, P. (1999). Alternative strategies for livestock-fish integration with emphasis on Asia. Ambio: A Journal of the Human Environment. 28(2): 118-124.

Mir, M.S., Naikoo, N.B., Amin, Z., Bhat, T.A., Nazir, A., Kanth, R.H. and Rehman, U. (2022). Integrated Farming System: A Tool for Doubling Farmer's Income. Journal of Experimental Agriculture International. 44(3): 47-56.

Mitra, A., Majumder, D., Mishra, M. and Sarkar, S. (2021). Poultry Farming: Prospects and Impediments in India. Saudi Journal of Humanities and Social Sciences. 6(6): 193-198.

Morrison, P.S., Murray, W.E. and Ngidang, D. (2006). Promoting indigenous entrepreneurship through small-scale contract farming: the poultry sector in Sarawak, Malaysia. Singapore Journal of Tropical Geography. 27(2): 191-206.

Mottet, A. and Tempio, G. (2017). Global poultry production: current state and future outlook and challenges. World's Poultry Science Journal, 73(2), 245-256.

Mridha, M.A.R., Hossain, M.A., Shah, A.A., Nahiduzzaman, M. and Uddin, M.S. (2017). Effects of supplementary feeds with different protein levels on growth and economic performances of Nile tilapia (Oreochromis niloticus) cultured in a rainfed rice-fish ecosystem. Journal of Applied Aquaculture. 29(2), 152-166.

Nageswaran, M., Selvaganapathy, E., Subbiah, V.R. and Nair, S. (2009). Demonstration and replication of integrated farming systems at Chidambaram. Report of MS Swaminathan Research Foundation (MSSRF), Chennai, pp. 16-53.

Nair, P.K.R. (2012). Integrated Farming Systems: Opportunities and Challenges. Agroforestry Systems, 86(3), 337-341.

Ngambeki, D.S., Deuson, R.R. and Preckel, P.V. (1992). Integrating livestock into farming systems in northern Cameroon. Agricultural Systems, 38(3), 319-338.

Odinwa, A.B., Emah, G. and Albert, C. (2016). Analysis of Rural Farmers Participation in Yam Production in Ogba/Egbema/Ndoni Local Government area of Rivers State. International Journal of Agriculture and Earth Science, 2(6): 21-25.

Odinwa, A.B. and Amugo, N.M. (2022). Women Participation in Poultry Farming in Selected Local Government Areas of Rivers State, Nigeria. Academic Journal of Agricultural and Horticultural Research, 2(1), 1-12.

Okigbo, B.N. (1995). Major farming systems of the lowland Savanna of sub-Saharan Africa and the potential for improvement.

Pandey, P.R., Sharma, H.O., Gupta, J.K., Mishra, P. and Chaurasiya, R. (2018). Integrated farming system: Only way to increase farmer's Income in a sustainable manner. Journal of Pharmacognosy and Phytochemistry, 12(1), 210-214.

Paramesh, V., Chakurkar, E.B., Bhagat, T., Sreekanth, G.B., Kumar, H.B.C., Rajkumar, S., Gokuldas, P.P., Mahajan, G.R., Manohara, K.K. and Ravisankar, N. (2021). Impact of integrated farming system on residue recycling, nutrient budgeting and soil health. Indian Journal of Agricultural Sciences, 91(1), 44–48.

Paul, I., Mandal, L. and Datta, S. (2018). Feasibility study of the small-scale poultry-cum-fish farming in homestead ponds for empowerment of rural women: A case study in the new alluvial zone of west Bengal, India. International Journal of Fisheries and Aquatic Studies, 6(3), 118-122.

Place, F., Ochieng, J., Wangila, J. and Barrett, C.B. (2019). Factors affecting input use in agricultural production systems in Western Kenya. Food Policy, 89, 101753.

Pudadera Jr., B.J., Corre, K.G., Coniza, E. and Taleon, G.A. (1986). Integrated farming of broiler chickens with fish and shrimp in brackish water ponds. In: The First Asian Fisheries Forum. Proceedings of the First Asian Fisheries Forum, 26-31 May 1986, Manila, Philippines (pp. 141-144). Asian Fisheries Society.

Radhamani, S., Balasubramanian, A., Ramamootthy, K. and Geetalakshmi, V. (2003). Sustainable integrated farming systems for drylands – A review. Agricultural Reviews, 24(3), 204-210.

Ram, S. and Das, S. (2020). Integrated Farming System: A Boon to Small and Marginal Farmers. Agriculture Update, 15(2), 190-195.

Ramanathan, K., Sangeeviraman, V., Chandrahasan, P., Chaudhary, B.N. and Ramachandra, S.S. (2020). Integration of fish culture and poultry rearing in transplanted rice for nutritional security in smallholder farms. Scientific Reports, 10(1), 1-7.

Ravisankar, N., Singh, P., Mishra, R.P., Prusty, A.K., Shamim, M., Singh, R., Tripathi, D. and Mohan, B. (2016). Annual Report of AICRP on Integrated farming system.

Rezaei, M., Yngvesson, J., Gunnarsson, S., Jönsson, L. and Wallenbeck, A. (2018). Feed efficiency, growth performance, and carcass characteristics of a fast-and a slower-growing broiler hybrid fed low-or high-protein organic diets. Organic Agriculture, 8(2), 121-128.

Roiter, L.M., Eremeeva, N.A. and Roiter, Y.S. (2020). Development of organic poultry farming based on environmental engineering. In: IOP Conference Series: Materials Science and Engineering (Vol. 919, No. 2, p. 022022). IOP Publishing. 10.1088/1757-899x/919/2/022022

Saha, B., Bhowmick, S., Patra, B. C. and Debnath, D. (2019). Integrated fish poultry farming: a sustainable approach towards nutritional and economic security in India. Aquaculture International, 27(1), 31-46.

Sanjeev, K., Singh, S.S. and Dey, A. (2011). Integrated farming systems for Eastern India. Indian Journal of Agronomy, 56(4), 297-304.

Sharma, S.K., Nainwal, G.C., Kumar, P. and Ujjania, N.C. (2016). Integrated fish-cum poultry-cum-horticulture farming system: an approach to up lift the socio-economic status of fish farmers. Progressive Research – An International Journal, 11(1), 411-414.

Shettar, D. and Rajeshwari, M. (2015). A study on issues and challenges of women empowerment in India. IOSR Journal of Business and Management, 17(4), 13-19.

Singh, J.P. and Ravisankar, N. (2015). Integrated farming systems for sustainable agricultural growth: Strategy and experience from research. In: Proceedings of National Seminar on "Integrated Farming Systems for Sustainable Agriculture and Enhancement of Rural Livelihoods, 13-14.

Singh, N.P. and Bhatt, N. (2021). Organic Poultry Production Management. Vigyan Varta, 2(3), 1-4.

Singh, R.P. and Ratan. (2009). Farming system approach for growth in Indian Agriculture. Lead paper in: National seminar on Enhancing efficiency of Extension for sustainable agriculture and livestock production, Dec 29- 30, Indian Veterinary Research Institute, Izatnagar.

Sinha, A. and Dey, A. (2021). Role of Integrated Farming System for Sustainable Agriculture. International Journal of Scientific and Research Publications, 11(6), 1-7.

Souillard, R., Répérant, J.M., Experton, C., Huneau-Salaun, A., Coton, J., Balaine, L. and Le Bouquin, S. (2019). Husbandry practices, health, and welfare status of organic broilers in France. Animals, 9(3), 97.

Thamrongwarangkul, A. (2001). Annual report on sustainable community development for good livelihoods and environmental project. KhonKaen University, Thailand.

Thangamani, S. and Muthuselvi, S. (2013). A study on women empowerment through self-help groups with special reference to Mettupalayam Taluk in Coimbatore District. Journal of Business and Management, 8(6), 17-24.

Tipraqsa, P. (2006). Opportunities and constraints of integrated farming system in Northeast Thailand. A case study of the Huai Nong Ian catchment, Khon Kaen province. Cuvillier Verlag.

Tipraqsa, P., Craswell, E.T., Noble, A.D. and Schmidt-Vogt, D. (2007). Resource integration for multiple benefits: Multi-functionality of integrated farming systems in Northeast Thailand. Agricultural Systems, 94(3), 694-703.

Tokrisna, R. (1995). Integration of agriculture, livestock and fish farming in Thailand. RAOS/CTA/FAO Seininar Proc, pp. 245-263.

van Noordwijk, M., Minang, P.A. and Freeman, O.E. (2018). Innovations in smallholder farming systems for improved livelihoods in the tropics. Current Opinion in Environmental Sustainability, 34, 23-28.

Verma, S.R., Kumar, P., Kumar, V. and Singh, S. (2017). Integrated fish and poultry farming: A means of improving resource use efficiency and economic feasibility. Journal of Entomology and Zoology Studies, 5(5), 898-901.

Yadav, S.P.S., Ghimire, N.P., Yadav, B. and Paudel, P. (2022). Key requirements, status, possibilities, consumer perceptions, and barriers of organic poultry farming: A review. Fundamental and Applied Agriculture, 7(2), 150-167.

Yun, C., Hu, C., Ma, Y., Xu, X., Hu, Z. and Zhou, Z. (2016). A review of integration strategies for sustainable aquaculture and poultry production. Frontiers in Marine Science, 3, 232.

Zaniboni-Filho, E., Pedron, J.D.S. and Ribolli, J. (2018). Opportunities and challenges for fish culture in Brazilian reservoirs: a review. Acta Limnologica Brasiliensia, 30, e302.

7

Integrated Fish and Pig Farming – A Prospective Concept for An Entrepreneurial Venture

Halpati Reena Prakashbhai[1], Krupesh S. Sawant[1] and Banlam Jingshai Marbaniang[2]

[1]Division of Fish Nutrition, Biochemistry and Physiology, ICAR-Central Institute of Fisheries Education, Versova, Mumbai, Maharashtra
[2]Division of Aquaculture, ICAR-Central Institute of Fisheries Education Versova, Mumbai, Maharashtra

Abstract

In this chapter we will discuss various aspects of fish and pig farming as an integrated model and study its potential from various angles to signify its capacity in enhancing profitability and productivity on a farm. Several studies have proven that integrating these two units effectively with a proper management system can result in an overall better performance of the farm. The benefit-cost ratio achieved from this system can reach as high as 3.36 which is multiple times greater than farming either of the animals alone. This system has the capacity to produce over 6 tons of meat and fish per year which are enormous values which will ultimately result in a very high net profit. With the increasing demand for fish and pork, especially in the North-Eastern Region, it is highly advisable for farmers and entrepreneurs to adopt this system, excel in the industry and improve their livelihoods. The production costs of this system are quite fair because a significant amount is saved on fish feed, manure costs, and other expenses that are produced as by-products by the pigs and used as inputs by fish for plankton generation and feeding. Nevertheless, more advancements are required to tackle various constraints such as disease outbreaks, availability of quality fish seeds and piglets, and devising ways towards ethical practices. This will bring about better changes in the sector.

Keywords: Integrated model, Recycle, Resources, Sustainable and Profitability.

Introduction

The development of software in the sphere of information technology and agriculture currently make up the majority of the economy (Shashank *et al.*, 2006). The majority of people in our country live in rural areas, so efforts to improve the economy must be concentrated there. To make any kind of progress in rural areas, strategies should be developed that will primarily focus on some crucial points, including the sustainable use of natural resources and their conservation, increasing production and profitability by efficiently using the resources already available, and improving quality and productivity by reducing waste.

Water management has been given significant emphasis in the formulation plan, according to Shashank *et al.*, (2006), who claim that it is becoming a global concern. Water is essential to life, hence issues like the scarcity of freshwater, lake and river pollution, etc., are concerning. In order to prevent water misuse and waste, efficient water usage and recycling in farming techniques are crucial. Integrated fish farming is a fantastic technique and the best example of mixed farming that efficiently uses water and recycles it for the culture of fish and other animals. According to Pandey and Malik (2016), pigs, ducks, and chickens are a widely used approach in Asia that has produced the highest yields in integrated fish farming so far.

This method of farming is one of the key ecologically balanced sustainable technologies used in various forms mostly in East and South-East Asian nations. The technology combines the production of crops or livestock with fish polyculture. Farm waste recycling, which involves using livestock excrement to fertilise ponds, pond water for irrigation of crops, or pond water for animals to drink, is a crucial aspect of integrated fish farming. It is very beneficial to farmers because it increases production efficiency by requiring less manure and feed and, it also lessens the negative environmental effects of farming (Shashank *et al.*, 2006).

Aquaculture and livestock or crop farming can be combined to produce high-quality protein foods, utilise resources, recycle farm waste, create jobs, and advance the economy. China, Hungary, Germany, and Malaysia are the countries with the most advanced integrated fish farming practices (Shashank *et al.*, 2006). Aquaculture systems following integrated fish farming models are recognized as sustainable.

By diversifying crops, integrated animal-fish farming may be a suitable strategy for maximizing profits from a small land area with limited investment (Jhingran, 1986; Mathias *et al*, 1997; Korikantimath and Manjunath, 2008).

One of the greatest ways to increase the production of animal and plant proteins is to adopt integrated agricultural systems, which make sustainable use of waste, water, and land resources (Sarm *et al.*, 2022). Nothing is wasted in this system, and ecological balance is preserved efficiently which is the prime motto towards sustainable development in order to sustain life and make no compromise for the generations to come.

According to Shashank *et al.* (2006), integrated fish farming may also be referred to as the sequential linking of two or more culture techniques for effective resource utilisation. It is the simultaneous growing of fish or shellfish along with other culture systems and when fish becomes its main component, it is referred to as integrated fish farming. The principle of integrated fish farming is the idea that "*there is no waste*", only misplaced resources that can be turned into valuable components of other products. In integrated fish farming practice trash or by-products from one system are successfully recycled and this style delivers excellent resource efficiency (Sarm *et al.*, 2022).

The basic idea behind erecting an integrated model of pig and fish is to utilise the output of one system as the input of the other which will eventually create better waste utilisation through which the farm income will greatly improve (Strasbourg, 2013). By adhering to the following principles, integrated fish farming acts as an example of sustainable food production:

a) The waste products of one biological system act as food for a different one.

b) The coexistence of fish and plants creates a polyculture, which boosts diversity and produces a variety of goods.

c) Through biological filtration and recirculation, water is reused.

d) The local economy is boosted and healthy food is made available.

Recycling organic wastes for fish culture accomplishes two goals at once: it benefits the environment and the economy. With a focus on using animal manures as fertiliser and nutrients to encourage the growth of natural feed in fish ponds, the integration of livestock with fisheries and aquaculture has recently attracted a lot of attention (Delmendo, 1980; Wohlfarth, 1979).

With potential proportions of 72-79 percent nitrogen, 61-87 percent phosphorus, and 82-92 percent potassium, the significant amount of nutrients (N–P–K) present in the animal feed being recovered in the manure serves as potential fertilisers in fish ponds to produce natural food known as plankton, which is a proteinaceous food item making up the basis of natural feeding.

Recent studies have shown that when animal manures are correctly applied to fish polyculture systems, significant fish output can be obtained (Shoko, 2011). Natural fish production relies on the recycling of animal manure and dung in aquaculture ponds, which promotes sustainable aquaculture and lowers costs for additional feeds and fertilisers.

Integrated Farming of Fish and Pig

Pigs can fit into a variety of management systems due to their vast range of food acceptance. Since pigs eat practically everything, feeding them is made easy for the farmer. Pigs are given a variety of foods, including household scraps, industrial or agricultural waste products, and offal from slaughterhouses, artificial feed, etc. Fish and meat are generated at a cheaper cost of feeding than other integration models, hence pig and fish integration, according to Kumar and Mishra (2011), is a preferable method. Pigs consume voraciously and can transform agricultural waste into delicious, very nutrient-dense meat. Among all domesticated animals, pigs are one of the most productive and best feed converters. A sow can have 16 piglets in a year if the generational gap is low. Pig farming offers rural residents work opportunities and significantly enhances their standard of living. Since the yield will come from two crops simultaneously, effective waste utilisation will occur, integrating fish and pig farming may lead to greater and better profitability. Pig excrement can be used as manure in fish and pig farming to raise fish in ponds naturally, while in pig rearing alone, the same waste may go to waste and be left to rot (Chakrabarty *et al*, 2014).

Fig. 1: Integrated Fish and Pig Farm in Jaintia Hills, Meghalaya
(*Picture Courtesy*: Shri. Pormi Suting, Farmer)

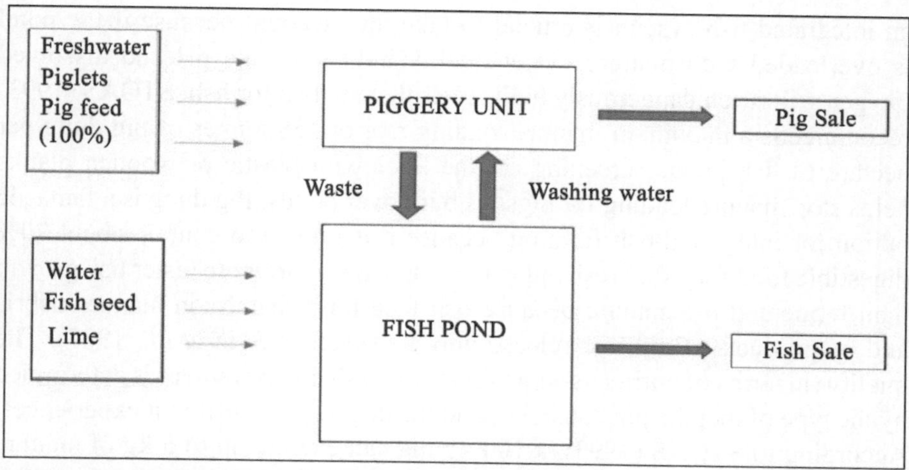

Fig. 2: Fish cum Pig Farming Model (Pandey and Malik, 2016)

Types of Pig Farming Systems

According to Panda *et al.* (2015), pig and fish farming are integrated in the following two ways:

i. **Direct Integration** – In this method, the pig pens are constructed on the pond embankment, and the excrement from the pigs is dumped directly into the fish pond, which eliminates the need for manual application and reduces labour requirements. However, uncontrolled discharge has the potential to result in high organic loads when done on days when manuring is not necessary. Eventually, this may result in the growth of excess algal matter, also known as algal blooms. To address this issue, a temporary waste storage facility is necessary.

ii. **Indirect Integration** – Pig dung is held in collection tanks and applied to the fish pond in the required quantity as and when necessary, which is the fundamental advantage of indirect integration in pig-fish farming. Dung collection tanks can be installed in any appropriate locations on the fish farm. The main drawback is the extra work and labour required for manual application.

Utilisation of Pig Manure

Panda *et al.* (2016) pointed out that modern concentrated animal farms produce a lot of manure as a by-product which contains considerable levels of both dissolved and undissolved organic material which can be potentially used in fish ponds or plant cultures for production. The dung loading rate of an integrated fish system is crucial to take into account because if the pond is overloaded with manure, water quality indicators like pH and dissolved oxygen will reach dangerously high levels that are bad for fish. STOAS (1993) recommends a maximum manure loading rate of 150 tonnes of pig dung per hectare of fish pond. Screening off the floor with plastic or wooden planks helps stop manure loading for pigsties built over ponds. Pig dung is a fantastic option for integrated fish farming because it is known to contain about 70% digestible food for fish. Fresh pig manure appears to promote faster fish growth than fermented pig manure because fish feed immediately on manure debris and by-products of nutrient release into the system (Yejin *et al.*, 1987). The quality (in terms of nutrient composition) of fresh pig excrement is determined by the type of diet the pig consumes and the degree of digestion it experiences. According to STOAS (1993), a 100 kg pig can excrete up to 5 kg of manure per day. Fresh pig dung contains about 18% organic matter, whereas nitrogen, P_2O_5, and K_2O account for 0.8%, 0.4%, and 0.3% of the total composition, respectively. There is a lot of natural food in a fish pond receiving pig waste, and the fish attain marketable size in a few months.

The steps of the manure usage process are illustrated as follows:

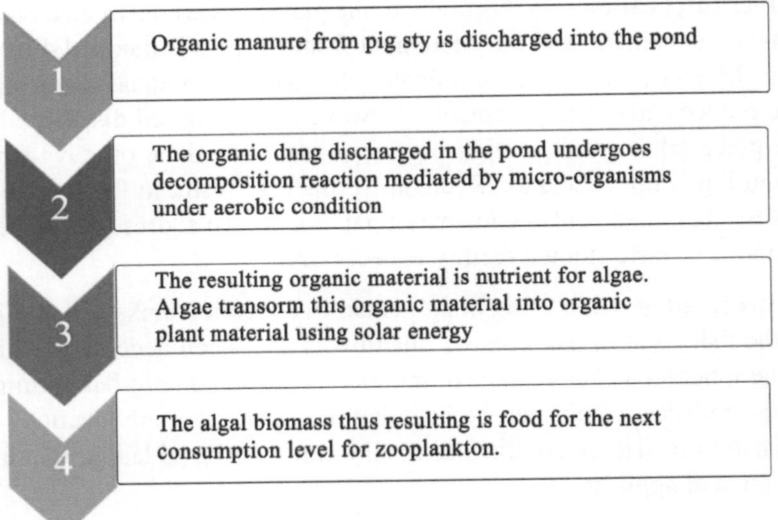

Fish naturally eat phytoplankton and zooplankton, which combinedly make up plankton biomass. The enormous amount of pig manure that builds up in pig farms is only partially used in the fields; therefore, the surplus, which is rich in direct and indirect proteins, can be effectively used to boost the natural nutrient content of fish ponds. Manure-borne microorganisms that enter the water supply also serve as fish food.

Table 1: Proximate Composition of Pig Dung (Source: Pandey and Malik, 2016)

Nutrient	Pig fed with Azolla (%)	Pig fed with Pig Mash (%)
Nitrogen	1.88	1.78
Phosphorous	0.60	0.45
Potassium	0.35	0.40
Moisture	70.00	69.50

Pig Breeds

According to Panda *et al* (2016) the following factors should be taken into consideration when choosing pig breeds to farm alongside fish:

a) Disease-resistant,

b) Rapid rate of expansion,

c) Meat of high quality,

c) Early adulthood,

d) Successful breeding. } If piglet production is a part of the practice.

In a pigsty close to the fish pond, exotic breeds such as White Yorkshire, Landrace, Hampshire, and Berkshire are raised. The Hampshire type is chosen for pig-cum-fish farming because it can produce 6-12 piglets in a litter, with an average growth of 60-70 kg in 6 months and a maximum growth of 110 kg in 6 months. It also reaches maturity in 6-8 months of growth. Thus, it is possible to raise two crops of this breed in a single fish crop in a single year. 40 to 50 pigs are enough to fertilise 1 ha of water spread area; hence 40 to 50 pigs are raised per ha of water spread area during the course of a 6-month fish culture period. Therefore, 80 to 100 pigs/ha of water spread area are needed for a one-year fish culture period. Weaned piglets between two and three months old are taken to the pig pen to be raised in this integrated system.

Following are the most popular pig breeds in culture systems:

1) Large and Middle white Yorkshire,

2) Berkshire,

3) Hampshire,

4) Duroc,

5) Hereford,

6) Landrace,

7) Chesterwhite

8) Tamworth
9) Crosses between exotic and indigenous pigs and pure local breeds are also available, for example, Ghungroo.

According to Chakrabarty *et al.* (2014), Large and Middle White Yorkshire and their crosses are best adapted to Indian conditions. For a long time, domesticated indigenous pigs served as the foundation for pig production but there are smaller in size and has a slower growth rate. Therefore, modern pig production in rural areas is based on improved breeds that have been graded up.

The following are the imported pig breeds raised in India, according to the Department of Animal Husbandry and Dairy, Government of India –

A. Large White Yorkshire

a) The most often used exotic species in India,
b) Its body colour is uniform white with sporadic black coloured spots,
c) Medium – length snout, erect ears, and dish-shaped face,
d) Outstanding breed for cross-breeding,
e) Successful breeders,
f) 300-400 kg mature boar, and
g) 230–320 kg for mature sow

B. Landrace

a) Long physique, huge drooping ears, and a long snout; white with a black skin patch; prolific breeding; and effective feed utilisation,
b) The quality of its carcass is equivalent to that of Yorkshire,
c) A superior breed for interbreeding,
d) Boars mature at 270 to 360 kg, and
e) Mature sows weigh 200 to 320 kg.

C. Middle White Yorkshire

a) Used in several regions of India,
b) Grows quickly and has a high dressing percentage;
c) Less common than huge white Yorkshire;

d) 250-340 kg males, and

e) 180–270 kg females

D. Niang Megha

a) Native to the state of Meghalaya,

b) Colour – Black with white star-shaped patches on the forehead and occasionally the shock joint.

c) Visible long, tapering snout, part-white nose, long, uniform bristles elsewhere on the midline, short ears, straight top line in males; concave in females.

E. Agonda Goan

a) The Agonda pig which was purified in the Old Goa complex is capable of adjusting to hot, muggy weather because it was carefully purified in regulated laboratory conditions. It belongs to a single, indigenous breed, is disease resistant, has a low fat content that consumers prefer in meat, and matures early.

b) Colour – Black, with a few animals having white patches on their faces and legs.

c) Visible characteristics include a small body size, medium and rough bristles, a concave top line, and a good adaptation to the local coastal environment.

F. Tenyi Vo

a) The Native Naga pig, or Tenyi Vo (literally, "Pig from Angami") as it is referred to locally, may be found in Chakesang, Mao, Tuensang, and Angami districts of Nagaland, situated in the extreme north-eastern corner of India along the Myanmar border.

b) Black in colour.

c) Visible characteristics – These animals have pot bellies, sagging backs, pendulous bellies that touch the ground in females, and straight tails that culminate with white markings that extend to the hock joint.

G. Nicobari

a) The tribes do not practise any systematic or scientific management because Nicobari pigs are semi-feral by nature. Due to this, tribal people have no knowledge of farming practises, and pigs were raised using a free-range approach. Pigs aren't typically raised for commercial purposes.

b) Skin tone – Fawn, black, grey, brown, and blackish brown.

c) Visible Characteristic – The facial profile ranged from flat to concave, and the back had a noticeable bristle crest (mane) that extended from the mid-head/shoulder to the base of the tail. The tail's distinguishing quality is that it does not curl. They can run quickly.

H. Doom

a) This breed is indigenous to the state of Assam

b) Colour: These are all black.

c) Visible Characteristics: The top line is straight, and the lengthy bristles reach the thoraco-lumber region.

I. Zovawk

a) Indigenous pig of Mizoram

b) Colour: Black with white spots on the belly, a white spot on the forehead, and white boots. The Zovawk pig is recognised by its long, midline bristles.

c) Visible Characteristic: The Zovawk pig's concave upper line is another distinguishing feature and have a pot belly.

J. Ghurrah

a) This breed is native to the state of Uttar Pradesh

b) Colour: They are medium-sized, black-coloured pigs with a flat belly, an angular body, and a long, straight snout.

c) Visible Characteristics: Legs are white below the hock joint. From the neck to the shoulders, there is a thick hairline. Short, leaf-shaped, vertically erected ears, triangular face, and extended head.

K. Purnea

a) The medium-sized black pig known as Purnea is found in the Bihar districts of Purnea and Katihar as well as nearby parts of the Sahibganj district of Jharkhand.

b) These pigs have pot bellies and compact bodies.

c) White markings on the lower limbs are, however, only occasionally seen in animals.

d) The animal has a thick line of bristles running from its neck to its shoulders, giving it a wild appearance.

e) The animals are distinguished by their round faces, short, conical ears that are erect, and small, thick snouts that are somewhat concave.

f) In mature animals, the skin is thick and has neck folds.

g) The average adult weighs between 41 and 50 kg.

h) At birth, a litter can range in size from 4 to 6. These pigs have a vicious disposition.

i) Black in colour.

j) Visible Characteristics: These pigs are known for their pot belly and compact bodies. White markings at the lower limbs are, however, only occasionally seen in animals. The animal's topline, which is thickly bristled from the neck to the shoulders, gives it a wild appearance. Animals are distinguished by their round faces, short, conical ears that are erect, and small, thick snouts that are somewhat concave. In mature animals, the skin is thick and has neck folds.

L. Ghungroo

Because of its high prolificacy and capacity to persist in low-input systems, the indigenous pig strain known as Ghungroo, which was initially discovered in North Bengal, is well-preferred by the locals. This breed or strain uses food scraps and agricultural by-products to produce high-quality meat. With a litter size of 6–12 piglets, Ghungroo is primarily black in colour and has the classic "Bull dog" face appearance. Individual piglets weigh about 1.0 kg at birth and 7.0–10.0 kg when they are weaned. Both sexes are quite docile and simple to deal with. They are kept in the breeding tract via a scavenging mechanism and primarily serve as insurance for rainfed agriculture.

Ghungroo pigs are raised intensively at the National Research Centre on Pig in Rani, Guwahati, utilising conventional breeding, feeding, and management practices. The productivity and reproductive effectiveness of this native strain are excellent, and their genetic potential for use in upcoming breeding programmes is now being assessed. Some of the chosen sows gave birth to litter sizes of 17 piglets, which were larger than those of other native strains of pigs kept at the Institute farm.

Advantages of Fish cum Pig Farming (Tripathi and Sharma, 2001 and Singh, 2019)

- Some of the fish feeds directly on the pig excrement, which provides 70% digestible food for the fish.
- Pig dung is a good fertiliser that aids in the growth of fish food, such as phytoplankton and zooplankton, because it includes 69 to 71 per cent moisture, 1.36 to 2.0 per cent nitrogen, and 0.36 to 0.39 per cent phosphorus when fed to pigs with pig mash having 16 to 17 per cent protein.
- There is no need for supplemental feed, which typically makes approximately 60% of the entire cost of input in conventional fish rearing.
- Pond water is utilised for washing the pigsties and bathing the pigs, and the pond dikes give room for the construction of animal housing units.
- It has particular significance in some regions since it can raise the socio-economic standing of rural communities in need, e.g., Pork has a very high demand in north-eastern region.
- Fish and pig meat can both be produced simultaneously in the same area.
- Fish ponds can provide the water needed for pig husbandry exercises.
- A portion of the pig feed can be made from crops grown in the fish pond embankment.
- On the pond's banks, horticultural crops like papaya and banana can be planted.
- The pond muck that accumulates at the pond bottom as a result of continual pig dung application can be utilised as fertiliser for growing vegetables and other crops as well as cattle feed, ensuring significant profit with little effort.
- Piggery enterprises do not need additional land.
- The pond embankments are used to cultivate grass and the cattle fodder needed for pigs.
- This model leads to high animal protein production per unit area.

Pig Sty Construction

Nnaji (2008) asserts that there are two different ways to build a pig sty –

i. On the dikes next to the fishpond, where the dung can be easily washed away and discharged into the pond. Concrete or another impermeable substance that can store pig excrement and urine should be used for the floor, which should slope towards the pond. The dung and urine must be transported into the pond via a pipe. When manure loading must stop, a diversion canal can be built to direct surplus manure into a compost pit.

ii. Over the pond, in which the dung falls directly into the body of water. The floor needs to be made of wooden planks or other materials that are spaced apart just enough to prevent the pigs' feet from slipping in and hurting them and permit dung and urine to fall directly into the pond.

For each pig, the sty should have a floor area of 1 to 1.5 m^2. To lower the cost of the system, pigsty construction can be done with locally available materials (bamboo, mud blocks, dry grass, etc.). A piggery unit with the dimensions 10 x 12 square feet is built, according to Pandey and Malik (2016), to raise pigs in integrated systems with appropriate provisions for feeding and drinking troughs. To enable automatic waste disposal into the pond, which can be done on a daily basis, a PVC pipe with a diameter of 4 inches and a length of 12 metres is fitted from the drainage outlet of the piggery unit and connected to the pond. To raise pigs in this type of method, weaned piglets aged 2 ½ months are purchased. Before stocking, male piglets should be castrated.

Infrastructures and Facilities

With food and drinking troughs, the pig pen should have good ventilation. The primary requirement for the construction of a pig house is adequate areas with all necessary amenities. Since pigs lack a heat regulation system, consideration should be given to this issue while building a pig home. According to Panda *et al*. (2016), a pig sty needs 1-1.5 m^2/pig of floor space, and 3–4 m^2/pig for pigs weighing 70-90 kg (Kumar and Mishra, 2011). The sty's height should range from 1.7 to 2.0 metres. The concrete wall should be between 1.0 and 1.2 metres high, and iron netting is installed over this concrete wall at a height of 0.5 to 1.0 metres. This iron netting aids in the pig sty's ventilation, or aeration, which keeps the pig house dry. An enclosed run is connected to the pig sty facing the pond to make sure the pigs have access to ample fresh air, sunlight, exercise, and space to relieve themselves. The pig house is kept clean by having feeding and drinking troughs integrated into the run (30 cm/pig) for convenient access. The fish pond is connected to the soak pit with a drainage pipe in the event of a direct integration system, and this soak pit is manually controlled as needed to discharge. The floor of the sty and the run is concreted with drainage facilities to a soak pit that has built-in shutter facilities. The ground has a slope towards the fish pond. This soak pit for pig sty wastes helps to store the pig sty wastes when the application of pig dung to the fish pond is stopped by closing the shutter. The indirect system of integration transports the pig sty wastes from the soak pit to the site of application in the fish pond as and when necessary, in the required quantity, as opposed to the direct system of integration, which connects the floor of the pig sty to a soak pit where the pig sty wastes are stored.

Housing

Pig farming requires well-protected housing because pigs are quite vulnerable to unfavourable climatic conditions, according to Chakrabarti *et al.* (2014). It must be adequately shielded from the sun's rays, the elements, and the wind. Pigs have a tendency to root, therefore the walls and floor should be sturdy enough to handle this. A wallowing tank may be put in the pig house to provide thermal comfort on hot summer days. Pigs can be housed in either an indoor system or an open-air system. These two approaches can be combined, as it is very beneficial for pig raising.

Stocking Density and Floor Space Requirement

Delmendo (1980) demonstrated that in mainland China it was typical to incorporate 30 pigs across 1 ha; Little and Muir (1987) reported the use of 60 pigs/ha; 70 pigs were integrated across a 1 ha fish pond that was stocked with approximately 20,000 Tilapia fingerlings, with an average weight of 30g, according to Tokrishna (1995). Kumar and Mishra (2011) claim that 40-45 pigs are sufficient to fertilise a 1 ha pond and that a single pig excretes about 500-600 kg of dung per year. According to Maguswi (1995), up to 100 pigs can be integrated across a 1 ha fish pond, which means that one pig weighing 50 kg can provide enough manure for a fish pond measuring 100 m^2. The pigs can be stored as weaners when they weigh an average of 10 kg; after nine months, they can weigh a maximum of 80 kg.

Procuring Piglets

Piglets should be purchased from trustworthy and verified breeders or from veterinary pig farms. Since the quality in terms of the piglets' health state is not guaranteed, buying piglets from local markets or unidentified people may result in a loss for the customer. Piglets intended for meat production should be castrated by expert veterinarian, and the appropriate immunisations should have been given at various times.

Breeding of Pigs

According to Chakrabarti *et al.* (2014), pigs can be bred at 8 to 9 months of age or 80 to 90 kg of body weight, and the boar-to-sow ratio should be 1:20. A sow can typically have three litters in every two years. Pigs have a 21-day oestrus cycle and the ideal time for mating is 24 hours following the commencement of heat, with a second mating occurring 8 to 12 hours later. Pigs go through gestation that extends to 114 days.

Care of Newly Born Piglets

A sow typically has 8 to 12 piglets every farrowing. When the piglets reach a body weight of 10 to 12 kg at 6 to 8 weeks of age, they are weaned (Chakrabarti *et al*, 2014). According to Ogbonna (2015), the following steps should be taken to prevent piglet mortality –

a) Extra heat should be made available,

b) Within 45 minutes of birth, the piglets should sip colostrum,

c) The navel cord should be cut 2 inches from the body and dipped into a weak iodine solution,

d) To avoid anaemia and poor health, the piglets should receive an iron injection within three days.

Feeding and Feeding Management

As long as the pigs accept a wide variety of foods, they can be fed various feed products. Pigs can be fed pig mash consisting of rice bran, rice polish, wheat bran, broken maize, mustard oil cake, fish meal, mineral mixture, and common salt, formulated at a ratio of 30: 15: 27: 10: 12: 4: 1: 1 (Pandey and Malik, 2016). The pigsty is cleaned before each feeding, which takes place between 7 AM and 5 PM daily. Piglets between the ages of 2½ and 4 months are fed 1.0 kg of feed, those between the ages of 4-6 months are fed 1.5 kg, and those between the ages of 6-8½ months are fed 2.0 kg.

Pigs are intensively raised in pig-cum-fish farming, and they are not allowed to leave the pigsty, as per the study by Panda *et al.* (2016). They are fed a balanced feed-pig mash in the sty itself at a rate of 1.4 to 1.5 kg per pig per day. The entire amount is split into 3 equal portions and fed to the pigs three times per day. They are fed green grasses, *Eichhornia*, and other things in addition to this balanced diet. Supply of enough drinking water is crucial, along with feed.

According to Nnaji (2008), pigs need food with a crude protein level of around 14%. Compound feed or a combination of compound feed and inexpensive feedstuffs like rice bran and kitchen scraps can be used to feed pigs. Since the pigs at this time do not even produce enough dung, the fish may also be given affordable additional feed like rice bran and wheat offal if the pigs are stocked as piglets or weaners. Pigs raised in intensive pig-cum-fish farming are fed a balanced pig ration in the home itself at a rate of 1.4-1.5 kg/pig/day. As a concentrate feed mixture, corn, groundnuts, wheat bran, fishmeal, and minerals are offered.

Feed for Breeders

An adult male pig or boar needs, on average, 2 kg of concentrate feed daily. Since it includes proteins, carbs, minerals, vitamins, lipids, and other nutrients, this concentrated blend is a healthy option. Every day, a sow in lactation needs at least 3 kg of feed while a pregnant sow needs 2.5 kg.

Managing Procedures

Pigs are the most economically advantageous, prolific breeders, and food converters among domestic animals. Pigs have a Feed Conversion Ratio (FCR) of 3 to 3.5 kg of feed per kg of live weight growth and are capable of quickly converting food waste, agricultural by-products, and industrial waste into high-quality animal meat.

The pigsty is constructed on the pond embankment in a way that a delivery channel allows the pig sty's waste washings to drain into the pond. A diversion channel is typically present to redirect the excreta away from the ponds since excessive excrement may cause algae blooms or other irregularities in the ponds. Pigsty washings are drained into the pond after dawn to prevent the loss of oxygen.

In order to keep the pigs healthy, the pigsty must be kept clean. For this reason, regular washing is a major priority. Additionally, water basins and feeders should be cleaned before each meal in order to eliminate uneaten feed and give the pig's fresh food. To prevent the spread of infections from the external environment through carrier organisms, the sty area should be cleaned and disinfected periodically as well.

Castration

Castration, performed by skilled veterinarians, is the surgical removal of the male reproductive organ, or testis, of piglets. Male domestic piglets are castrated primarily to avoid unpleasant boar taint and to lessen aggressive and sexual behaviour (Weiler, 2016). There are a couple of major reasons behind this practice, first being the energy expenditure during breeding season i.e., during breeding season, boars are highly sexually active and this results in maximum energy being utilised sexually which ultimately impacts on the growth of the pig; therefore, removal of testis diverts the pig's 100% energy into its growth. The second reason, as mentioned earlier is the boar taint which is emitted from the meat of a pig whose testis has not been removed, this creates a marketing issue wherein customers usually reject meat having boar taint.

Health Management

Although pigs are sturdy creatures, they are prone to illnesses such as swine fever, swine plague, and swine pox as well as parasite infestations like roundworms, tapeworms, and liver flukes. (Panda *et al.*, 2016). Pigs are kept safe from disease threats by maintaining clean and healthy conditions in the pigsties. Daily morning and evening cleanings of the pigsties are required. During the summer, the pigs should also be given a bath. Disinfectants like potash or lime should be used at least once a week to clean the pigsties and other equipment utilised in pig husbandry activities. Pigs should periodically be dewormed. The cultured stock of pigs should receive iron injections like Imferon to prevent anaemia. Animals in culture should be immunised against all viral diseases. Pigs fed on kitchen garbage have been observed to have vitamin and mineral deficiencies; thus, to meet the body's needs, external vitamins and minerals in the form of powder supplements should be added to the food or administered as injections. According to Godyn *et al.* (2020), pigs are more prone to heat stroke since they are sensitive to exposure to high air temperatures which causes behavioural and physiological reactions, which reduces their productivity. The techniques most frequently employed in pig facilities to lessen the impacts of heat stress include the evaporative pad cooling system, high pressure fogging, sprinkling, as well as rising air velocity and floor cooling.

Production and Harvest

The pig grows to an average live weight of 65 kg after reaching slaughter maturity during a 6-month growth period, at the point of which it is harvested.

A new batch is introduced to grow once the entire batch has been removed from the culture system. A hectare of water spread area can be predicted to produce 6000 kg or more of pig meat and roughly 6000 kg of fish as well.

Fish Farming Practices

The practice for raising fish in pig-cum-fish farming is integrated using a "composite fish culture method." The composite fish culture system can be described in detail as follows:

Composite Fish Culture

The purpose of this culture is to maximise the utilisation of fish ponds. Determine the sort of fish that is chosen for composite fish production by looking at its feeding habits, including whether it is a surface feeder, column feeder, bottom feeder, eater of aquatic plants, omnivorous, etc. All management techniques utilised in a composite fish culture system can be divided into three categories: pre-stocking management, stocking management, and post-stocking management.

In a 0.12 hectare area, 8000 advanced fingerlings of IMCs viz., *Catla catla*, *Labeo rohita*, and *Cirrhinus mrigala* can be stocked along with common carp, *Cyprinus carpio var. communis*, in a 4: 3: 2: 1 ratio. This fish farming method was also used by Pandey and Malik (2016). The appropriate pre-stocking, stocking, and post-stocking management procedures should be followed.

There are three different types of ponds in a scientifically designed fish farm, *viz*.

Nursery pond: The size of the pond should be between 100 and 500 square meters, and the water should be no deeper than 1.5 meters. This pond makes up 5% of the fish farm's total productive area.

Rearing pond: The area of the pond is from 500 to 1000 m^2, and the water's depth is between 1.5 and 2.0 m. 15% of the fish farm's total productive area is made up of this type of pond. It might also be used occasionally as a stocking pond.

Stocking pond: The stocking pond's size spans from 1000 to 20,000 m^2, and its water depth is between 2 and 2.5 m. This kind of pond makes up 60 to 70 percent of the fish farm's total productive area.

Bio pond: In addition to the ponds indicated above, several farms now have a particular type of pond called a bio pond. It serves as a sizable settling tank where biological filtration of agricultural water used for fish ponds takes place.

It may also be utilised as a stocking pond, depending on the need. A fish farm's overall productive area is 7–10% of the area occupied by this kind of pond.

Pond Preparation

1. Pre-stocking management

a) **Liming and fertilisation:** The pond is filled with water to a depth of 2 to 2.5 meters for liming and fertilisation. The pH of the water is then measured, and lime is added based on the pH result. After being limed for 7–10 days, pond water is fertilised. The pond is fertilised using both organic and inorganic fertilisers. Fish seed is then discharged into the pond water 7-15 days after application of inorganic fertiliser and evaluating the natural productivity. Nitrogen and phosphorus are abundant in pig excrement, as a result, this sort of integrated farming does not require the additional fertiliser described above. Ten to fifteen days before stocking the fish pond with fish seed, pig dung is applied at 1000 kg/ha and at a rate of 50 kg per ha each day after stocking as part of the indirect integration process.

b) **Aquatic weed removal:** Herbicides like 2, 4-D and gammaxene can be used to get rid of aquatic weed.

c) **Weed fish removal, unwanted organism removal, etc.:** Ponds can be cleaned of undesired fish by using mahua oil cake. Because the water's toxicity may last for up to 15 days, stocking should only be done once it has been determined that the water has entirely detoxified.

d) **Algal bloom control:** Controlling algal blooms can be done with the use of chemicals like diuron (0.3-0.5 mg/lit of water) and $CuSO_4$ (0.1-0.5 mg/lit of water).

e) **Removal of toxic gases, etc.:** Toxic gases and other consequences of the mud can be reduced by continuously dragging a rope along the bottom of the pond.

f) **Pond embankments:** should be fortified and raised high enough to prevent any potential fish access from neighbouring water after significant rain.

2. Stocking Management

a) **Selection of fish species**: There are numerous fish species that can be cultivated in this farming system. However, the species chosen for composite fish culture should have the following qualities:

i. Fast growth rates, good food conversion rates, acceptance of supplemental and natural food, ability to adapt to crowded conditions, and disease resistance.

ii. Capacity to endure shifting physical, chemical, and biological pond water conditions.

iii. A fair market price.

b) **Stocking of fingerlings:** At least 7000–8000 fingerlings per hectare must be stocked. You can introduce combinations of 3, 4, or 6 species into the pond.

c) **Fish species combination**: There are three possible species compositions: Catla: Rohu: Mrigal, Catla: Rohu: Mrigal: Common Carp, and Catla: Rohu: Mrigal: Common Carp, Silver carp, and Grass carp. Silver carp is a particularly favourable species to stock in bottom-feeding Mrigal with a focus on filter-feeding species, according to Kumar and Mishra (2011).

Table 2: Fish Species Combination in Fish cum Pig Farming (Source: Panda *et al.*, 2016)

Fish Species	**Species combination proportion (%)**		
	3 species	**4 species**	**6 species**
Catla catla	50	50	30
Labeo rohita	30	25	20
Cirrhinus mrigala	20	20	10
Cyprinus carpio	-	5	5
Hypothalmichthys molitrix	-	-	20
Ctenopharyngodon idella	-	-	15

3. Post Stocking Management

a) **Liming:** Liming helps maintain a steady pH in fish pond water. The natural production of the pond is increased as a result. Liming also benefits the wellbeing of the stock of farmed fish.

b) **Fertilisation:** Fertilisation boosts the pond's natural food supply. At the same time, fertilisation causes a variety of environmental issues, including decreased dissolved oxygen levels, phytoplankton blooms, increased NH_3 levels, etc. However, it is thought that manuring by itself can boost pond productivity by 75%.

c) **Water quality management:** Siltation is a problem in pig-cum-fish farming, which is a major contributor to turbidity in the water of the fish ponds. To investigate the problem, different areas of the fish pond should be treated with pig dung each day. Every year, following partial dewatering, the pond bottom muck needs to be removed once again. Dewatering and desilting must be done completely after three years for this integration.

d) **Phytoplankton bloom:** Phytoplankton bloom is the term used to describe the abrupt rise in population of a particular group of planktonic algae as a thick mass in water. It can be recognised by the pond's deep green, blue green, or reddish green water. This pond water's algal bloom is brought on by an excess of nutrients in the water. Therefore, the pig dung supply should be immediately turned off and the proper corrective action should be done if this problem occurs in the fish culture pond.

e) **Harvesting Management:** After four months, the fish may be harvested, and there may be two fish harvests over a nine-month cycle. The surplus fingerlings can be sold, while any present fingerlings can be collected and stocked in various ponds for the upcoming growing season (Nnaji, 2008). Fish yields from the pig-fish system can reach 11,250–12,500 kg/ha/yr. (Tokrishna, 1995), 40–80 kg fish per 100 m^2 (Maguswi, 1995), and 3–4 tonnes/ha after 12 months of culture without feeding (Kumar and Mishra, 2011).

In fact, the choice of when to harvest is left up to each farmer, and it is critical to sample the fish pond frequently (for instance, once a month) in order to track fish growth and choose the best time to sell.

Economics

Integration of fish and pig in a single model has been proven to be a very profitable venture for a farmer if management practices are carried out efficiently. Below is the input-output analysis of fish and pig integrated farming system according to the study carried out by Aziz *et al.* (2018) in Assam. The benefit-cost ratio of this model has been reported to be 3.16 (Bhagawati *et al.*, 2020), 3.36 (Kalita *et al.*, 2021) and 1.40 (Aziz *et al.*, 2018).

Table 3: Economics of Pig cum Fish Farming

S. No.	Particulars	Rate (₹ per unit)	Amount (₹ in lakh)
A	Fixed cost		
1	Construction	2.5	2.5
2	Electric installation & Electrification	0.5	0.5
3	Equipment	1.0	1.0
4	Miscellaneous	0.5	0.5

S. No.	Particulars	Rate (₹ per unit)	Amount (₹ in lakh)
5	Watchman shed	0.5	0.5
6	Fencing	0.5	0.5
	Total	**6.50**	**6.50**
B	**Variable cost**		
1	Lease amount	03/ year	03
2	Fertilisers and Chemicals	03 lump sum	03
3	Cost of seeds (Fish + Piglet)	25000 yearlings @ ₹8/- 50 piglet @ ₹4000 each	02 02
4	Cost of feed (Fish + Pig)	25 tonnes @ 40/kg 10 tonnes @ 20/kg	10 02
5	Fuel and electricity charges	0.5	0.5
6	Labour charges	2 workers @ 12500/- for 10 months	2.5
7	Miscellaneous	0.5	0.5
	Total		**25.50**
	Grand Total (A + B)		**32.00**

Table 4: Input-Output Analysis

Sl. No.	Particular(s)	Production (Tonnes)	Rate	Amount (₹ in Lakh)
1	Fish	25	1.2 lakh/tonne	30.00
2	Pig (50 no. × 60 kg)	3	5 Lakh/tonne	15.00
Total				**45.00**
Net Profit = Production – Input = 45.00 – 32.00 = ₹ 13.00 lakhs BCR = 1.40*				
*Aziz *et al.* (2018) conducted their study in 2 stations but in this chapter economics of just one station has been depicted for easy interpretation thereby, BCR is lower. (Actual BCR as per Aziz *et al.*, 2018 is 1.80)				

Case Study on Pig cum Fish Integrated Farming

A field trial on pig cum fish integrated farming at the Livestock Research Station in Mandira in Kamrup (R) district of Assam was undertaken in 2018 and 2020. According to Kalita *et al.* (2021), the outcome of the pig cum fish integrated farming was reflected positively. Pig and fish farming generated a net profit of ₹1,21,700.00 in 2018 and ₹1,49,200.00 in 2020, respectively. A significant net return was observed due to the greater market price of both fish and pork in 2020, Pig excrement serves as a great pond fertiliser, increasing the pond's biological productivity and, as a result, the amount of fish produced. The fish culture does not require supplemental feed, which saved approximately 60% of the total input cost which is usually incurred in traditional fish culture.

Pig excrement contributes to pond water fertilisation and helps to create fish food organisms like phytoplankton and zooplankton. In addition, certain fish, including common carp, consume pig manure as their sole source of nutrition. For the local small farmers, the economic returns are sufficient. From the two years of experimentation, the B: C ratio was 3.36:1.

Integration of Fish-Pig-Azolla

A comparison research was conducted by Pandey and Malik (2016) to evaluate the growth parameters of the fish-pig-azolla integration model. Azolla serves as an additional meal for the pigs in this approach. According to the findings, this integration strategy performs better in terms of growth compared to fish-pig integration alone. The fish-pig integration and fish-pig-azolla models were able to produce a total fish output of 4427 kg/ha and 5315 kg/ha, respectively, following a 12-month culture period. In contrast, the average pig weight growth after six months was 62.30 kg per pig in the first model and 64.45 kg in the latter.

Conclusion

Taking into account the above factual investigation, it can be stated that combining fish and pig farming can be a profitable and environmentally friendly strategy that not only can increase the farmers' revenue by boosting fish production but also can create jobs through the raising of pigs. Several studies and works done on this aspect have shown that fish cum pig integration is much more productive compared to fish farming or pig farming alone. This model is eco-friendly and sustainable in nature, thereby it does not pose any harm to the environment in the long run. The effective utilization of pig manure as pond fertiliser is a great example of it being a waste recycling system wherein all resources are judiciously used and recirculated. With the increasing price of feedstuffs for fish, this system directly reduces feed costs which constitute over 60% of the total farm investment making it a feed-independent farming system and ultimately providing a better economic return for the farmer.

References

Aziz, A., Joysowal, M., Mondal, A., Singh, S.M., Kumar, V. and Boda, S. (2018). Economic Analysis of Integrated Pig cum Fish Farming in Assam. International Journal of Advances in Agricultural Science and Technology, 5(3): 64-70.

Bhagawati, K., Tamuli, K.K. and Saharia, P.K. (2020). Integrated fish cum pig farming system for enhancing farm income in Assam. J. Krishi Vigyan, (Special Issue): 209-213.

Chakrabarti, A., Chandran, P.C., Gupta, J.J., Reena Kamal, Shanker Dayal and Rajni Kumari (2014). Pig cum Fish culture in Integrated Farming System. Krishisewa.com. Division of

Livestock & Fishery management, ICAR Research Complex for Eastern Region, ICAR Parisar, P.O. Bihar Veterinary College, Patna- 800014, Bihar. Accessed on 20/06/2022.

Delmendo, M.N. (1980). Review of integrated livestock-fowl-fish farming systems. In: Integrated agriculture-aquaculture farming systems: proceedings, ICLARM-SEARCA Conf. Integrated Agriculture-Aquaculture farming Systems, Manila, Philippines, 6-9 August 1979/edited by R.S.V. Pullin and Z.H. Shehadeh. Manila, Philippines: International Center for Living Aquatic Resources Management, 1980.

Godyń, D., Herbut, P., Angrecka, S. and Corrêa Vieira, F.M. (2020). Use of different cooling methods in pig facilities to alleviate the effects of heat stress – A review. Animals, 10(9), 1459.

Jhingran, V.G. (1986) Aquaculture of Indian major carps, in Aquaculture of Cyprinids (eds. R. Billard and J. Marcel), INRA, Paris, pp. 335-346.

Kalita, M., Sonowal, M., Brahma, J. and Boro, P. (2021). A study on pig cum fish integrated farming in Kamrup (R) district of Assam. Journal of Entomology and Zoology Studies. 9: 462-463.

Korikantimath, V.S. and Manjunath, B.L. (2008). Integrated farming systems for sustainability in agricultural production. Proceedings of national symposium on new paradigms in agronomic research. Indian Society of Agronomy. Navsari Agriculture University. Gujarat, 2008, pp. 79-281.

Kujur, N., Prasad, C.M., Singh, A.K. and Singh, S.K. (2005). Economic analysis of integrated pig-cum-fish farming in Jharkhand. Indian Journal of Animal Research, 39(1), 73-75.

Kumar, K. and Mishra, B. (2011). Integrated Fish Farming. In: Ayyappan, S., Moza U., Gopalakrishnan A., Meenakumari B., Jena J.K. and Pandey A.K. (eds.) Handbook of Fisheries and Aquaculture, 2nd edn., ICAR New Delhi, DKMA, pp. 439-448.

Little, D. and Muir, J. (1987). A guide to integrated warm water aquaculture. Stirling, Scotland, Institute of Aquaculture, University of Stirling, p. 238.

Lock, K. (1993). Fish farming in ponds in the tropics, Fish farming in ponds in the tropics, Vol. 1.

Maguswi, C.T. (1995). Mixed animal farming in Zambia. In: J.J. Symoens and J.C. Micha (Eds). The management of integrated freshwater agri-piscicultural ecosystems in tropical areas, CTA/Royal Academy of Overseas Sciences Seminar, pp. 438-442.

Mathias, J.A., Charles, A.T. and Baotong, H. (1997). Integrated Fish Farming. CRC Press.

Nnaji, J. (2008). Integrated Pig-Fish Farming. http://dx.doi.org/10.13140/RG.2.2.14999.68000

Panda, S., Panigrahy, K.K., Babu, L.K. and Kumar, A. (2015). Fish-cum-Pig Farming. Indian Farmer, 2(11): 835-854.

Pandey, N. and Malik, D.S. (2016). Integrated Fish Farming. pp. 190, Daya Publishing House, New Delhi.

Rajesh Singh (2019). Income generation and nutritional security through Livestock Based Integrated Farming System. Pashudhan Praharee. https://www.pashudhanpraharee.com. Accessed on: 20/06/2022.

Sarm, M., Brahma, J., Boro, P., Sonowal, M., Bharali, D., Deka, A. and Bhattacharyya, H.K. (2022). A comparative analysis of integrated fish-pig and fish-duck farming in lower Brahmaputra valley zone of Assam. The Pharma Innovation Journal, SP-11(5): 336-340.

Shashank, S., Prabjeet, S., Nitin, V. and Diniesh, K. (2006). Integrated fish farming – rationale and scope. College of Fisheries, GB Pant University of Agriculture and Technology, Pantnagar, Aquafind, Aquatic Fish Database, India.

Shoko, A.P., Getabu, A. and Mwayuli, G. (2011). Growth performance, yields and economic benefits of Nile tilapia, Oreochromis niloticus and Kales Brassica oleracea cultured under vegetable-fish culture integration. Tanzania Journal of Science, 37: 37-48.

STOAS (1993). Integrated fish farming in the tropics Book 1 Aquaculture volunteered in Thailand foundation for the development of Agricultural Education and training Wageniga Netherland. pp. 40- 42.

Strasbourg, L.K. (2013). Fish Feeding in Integrated Fish Farming. pp. 272, Random Exports, New Delhi.

Tokrisna, R (1995). Integration of Agriculture. Livestock and fish farming in Thailand. In: J.J Symoens and J.C Micha (Eds). RAOS/CTA/FAO Seminar Proc. 1995. Pp. 245-263.

Tripathi, S.D. and Sharma, B.K. (2001). Integrated fish-pig farming in India. FAO Fisheries Technical Paper, 54-56.

Weiler, U, Stefanski, V. and von Borell, E. (2016). Castration of pigs – Conflicting aims and possible solutions from an animal welfare point of view. Züchtungskunde. 88: 429-444.

Wohlfarth, G.W. and Schroeder, G.L. (1979). Use of manure in fish farming – A review. Agricultural Wastes, 1(4): 279-299.

Yejin, Y., Yun, Z., Dan, H. and Junhua, W. (1987). Preliminary studies on the effects of fresh and fermented pig manure on fish production. Proceedings of integrated fish farming systems. Fresh water Fisheries Research Centre of Chinese Academy Fishery of sciences. pp. 324-327.

8

Entrepreneurial Opportunities of Agro-Forestry Based Integrated Farming System

***Duvvada Sarath Kumar*[1]*, Mahua Banerjee*[1]*, Manisankar G*[1] *Ganesh Ch. Malik*[1] *and Kajal Sengupta*[2]**

[1]*Department of Agronomy, Palli Siksha Bhavana, Institute of Agriculture Visva Bharati, Sriniketan, Bolpur, West Bengal*
[2]*Division of Agronomy, Faculty of Integrated Rural Development and Management (IRDM), School of Agriculture and Rural Development, Ramakrishna Mission Vivekananda Educational and Research Institute [Deemed to-be-University], Narendrapur Campus, Kolkata, West Bengal*

Abstract

Ninety-nine million of India's 121 million agricultural holdings are small and marginal (87%). The slow growth of Indian agriculture is entirely attributable to the fact that small holdings are not well adapted to farm mechanisation and the agricultural community's dire economic circumstances and lack of resources. Therefore, there is a great need to create integrated agricultural system models that are appropriate to the area in order to handle the multi-dimensional issues that are now present in agriculture and related industries. The adoption of integrated agricultural systems is one of the most efficient ways to increase farm households' production, profitability, nutritional security, livelihood development, job creation, and sustainability. This is especially true for small and marginal farming communities, which currently make up more than 87% of all farm households. Due to India's declining productivity due to rising population demands combined with dwindling farm resource availability, agricultural research programs urgently need to reorient from an individualistic enterprise approach to a developed holistic approach of an integrated farming system. The idea of integrated farming systems research takes a variety of enterprises and resources (inputs) at the farm level into consideration when planning the production of crops, selecting cropping systems, and combining various enterprises to develop integrated farming systems with sustainable agriculture production systems. Similarly, integrating a two- or three-tier system of integrated farming with poultry farming, horticulture, beekeeping, dairy farming, fisheries, sheep and goat rearing, sericulture, and mushroom cultivation can result in a significant increase in the production of high-energy food without affecting the production of food grains. Since agroforestry systems promote biodiversity, they have various advantages for halting desertification and mitigating climate change. In addition to potentially reducing deforestation and forest degradation, agroforestry practices can assist in reducing the demand for resources in natural conservation areas. Recently, since they provided smallholder farmers with fresh business prospects, it has become more and more pertinent. Combining crops with forest trees can increase productivity and profitability for marginal and small farms. Therefore, it is evident that switching to associated businesses significantly increases the net return compared to producing arable land.

Keywords: Agro-Forestry, Entrepreneurial Opportunities and Integrated Farming System.

Introduction

To provide a good standard of living, food and eco-friendly natural resources are essential. It is commonly acknowledged that in the current situation, attention must be given to sustainable development, conservation, and management of natural resources if we are to ensure the future of food, living, and ecological security. By the end of the twenty-first century, it's predicted that there will be 8 billion people on the planet, which means that 120 million tonnes of food will be needed to feed everyone (Gill *et al.*, 2009). By the year 2050, there is expected to be a 60% global increase in food demand and a 100% increase in developing countries. It is crucial to preserve natural resources such as soil, water, and vegetation, to meet future demands and keep up with population growth. Several farming systems coexist throughout India based on the resources available, agricultural practices, and location-specific needs of the population to meet their needs for food, fodder, fuel, and fibre (Jayanti *et al.*, 2003). A complex network of interconnected components, such as soil, plants, animals, tools, energy, labour, money, and other inputs, makes up an integrated farming system (IFS). Farming families govern it and are, to varied degrees, affected by institutional, political, economic, and other active variables at the farm level (Barbieri and Valdivia, 2010). As there is no room for horizontal land expansion for agriculture because of the nation's constantly growing population and declining per capita land availability (Shekinah *et al.*, 2005). Only vertical growth is feasible by combining agricultural and related businesses that take up less space and time while providing farm families with reasonable returns. Integrating various agricultural businesses, including farming, backyard poultry, animal husbandry, fisheries, forestry, etc., has great potential for the agricultural economy and is a good way to boost a farm's income. (Ajijur *et al.*, 2017). These businesses help farmers supplement their income and contribute to year-round increases in family labour employment and good entrepreneurial opportunities (Figure 1). One of the alternative strategies for reducing climate change is an integrated farming system. We can recommend cropping systems for climate mitigation, including intercropping, multiple cropping, crop rotation with legumes, conservation farming, agroforestry, alternating with non-legumes, and integrated farming. Agriculture practices like the liberal application of artificial fertilisers and pesticides during the 20th century significantly increased output. Still, unfavourable environmental damage and rising operating costs sparked worries about the viability and sustainability of agriculture. Agroforestry is the practise of growing shrubs, trees, and annual crops together. It keeps it up by providing steady employment, shade, a reliable stream of food or revenue throughout the year, halts degradation, and preserves soil fertility. An alternative management technique to traditional agricultural methods, agroforestry increases the possibility for farm business success while meeting the standards for reduced environmental impact. Farmers from underdeveloped nations now have access to commercial opportunities in regional and global markets because to agroforestry. Despite the fact that agroforestry-related goods and services now have more access to markets, their importance for the growth of rural and business enterprise is still underappreciated. The lack of appreciation for the current agroforestry practises might be attributed to ignorance of the critical role that these practises presently play in local economies and the potential of these systems for future agricultural expansion. By integrating the agroforestry with the crop production can enhance the business opportunities which in turn enhance the living standards of farmers.

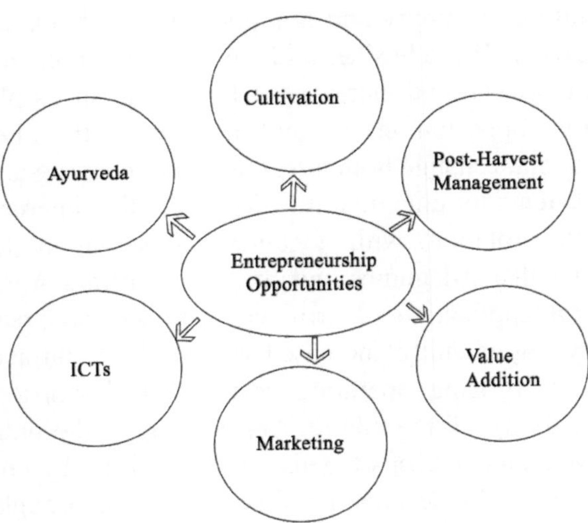

Fig. 1: Entrepreneurial Opportunities of Different Agroforestry Systems
[*Source*: Reddy, 2021]

Entrepreneurial Opportunities of Different Agroforestry Trees

Acacia mangium (also known as Mangium) is one of the most frequently utilised fast-growing tree species in plantation forestry operations in Asia and the Pacific. It is a significant source of wattle timber, used for veneer, furniture, boatbuilding, and construction (Krisnawati *et al*., 2011). It offers high-quality charcoal and can be used to make charcoal briquettes and synthetic carbon. The pulp is excellent for creating paper because it can be easily bleached to high brightness levels (Darus, 1993). The products from *Acacia* motivate farmers for better marketing and setting up enterprises. Babul (*Acacia nilotica*) is a versatile plant that produces a variety of goods, such as wood, firewood, bark gum, and fodder leaves. Construction and agricultural equipment are made from wood, including shafts, harrows, bullock cart yokes, ploughs, and wheels. The tree's branches provide excellent firewood. The wood produces premium charcoal. Wood is also used to make pulp and paper (Viswanath *et al*., 2000). For fishing floats, packing cases, and sword sheaths, *Ailanthus excelsa* wood is used. In the paper industry, pulp is used as a substitute for aspen and is made from debarked wood.

The leaves are considered extremely tasty, nutrient-rich protein feed for sheep and goats, which also increases milk production (Rajasugunasekar, 2014). *Alnus nepalensis* foliage is unsuitable for cattle and has a low to moderate value as fodder for sheep and goats. It is a valuable source of fuel and charcoal dries quickly, and burns efficiently (Sharma *et al*., 2007). *Anthocephalus cadamba* wood is widely used for planking, ceiling boards, packing boxes, light construction, carving, and turnery. Additionally, it can be used to make splints, matchboxes, and pencils. It is reportedly good for manufacturing wrapping and printing paper as well. Cattle are fed with fresh leaves. Flowers are a crucial component of the raw materials used to make attar which are Indian perfumes with a sandalwood foundation in which one of the essences is hydro-distilled (Vijayaraghavan, 2014). Neem (*Azadirachta indica*) tree twigs were used to make lumber, lubricant, soap, and lamp oil. Its wood is used to make toys, drums, cabinets, packing cases, cart axles, boards and panels, cabinets ornamental cuttings, helms, oars, cigar boxes, and agricultural implements, in addition to being used to build ships and boats. In many parts of the country, the leaves are still utilised as traditional animal feed since they are palatable to cattle and buffaloes. Most commonly, soap and toothpaste are made with its oil and its derivatives. Rheumatism, leprosy, and sprains can all be treated with oil externally, along with several chronic skin conditions, including ulcers (Subapriya and Nagini, 2005, Pankaj *et al*., 2011).

Some regions grow *Albizia lebbeck* primarily as cow, buffalo, and camel fodder. Sheep do not consume substantial amounts of saponin-containing pods, whereas cattle readily consume them. The white blossoms on the tree provide a sweet scent that attracts bees helping beekeepers obtain good quality honey. A reddish gum from the trunk is used to adulterate gum (Singh and Agarwal, 2018). Bamboo (*Bambusa vulgaris*) is often utilized in urban homes to create handicrafts, furniture, and decorative items. The paper industry in India uses a significant amount of bamboo produced. In agricultural operations various objects are utilised, including winnowers, planks, baskets, containers, ploughs, and a variety of others that were prepared by using bamboo (Hossain *et al*., 2016). The pulp yield and strength of *Casuarina equisetifolia* are sufficient for duplex paper and wrapping paper. House pillars, rafters, electric poles, tool handles, oars, wagon wheels, and mine props are all made from the wood of *Casuarina* spp. It is a promising agroforestry species for arid and semi-arid regions due to its high productivity and traits that improve soil fertility. It has a deep taproot, can resist cyclonic storms, and serve as a windbreak (Nicodemus, 2014). *Ceiba pentandra* is well suited for light plywood containers and is in high demand for matchwood. Additionally, it's utilised to load things like dugouts, cases, shingles, and brush handles. Silk cotton, also known as Indian Kapok, falls from the tree and is used for packing, cushioning, and other purposes. Mochras gum (Mochras is gum resin of '*Bombax ceiba* or *B. malabaricum*'. It is also known as '*Shalmali*' in Ayurveda) that emerges from the bark, has tremendous therapeutic significance (Orwa *et al*., 2009). *Dalbergia sissoo* is utilised for tool handles, marine and aviation grade plywood, decorative veneer, engraving, fine furniture, cabinets ornamental turnery, carving, and sporting goods. Tobacco pipes are made from root wood. It is well-known for producing doors and windows in the local economy. When other fodder forms are scarce, foliage serves as a last resort source of nutrition (Bijauliya *et al.*, 2017). Boxes, pallets, bins, scaffolds, and other items are made from green saw wood of *Eucalyptus tereticornis* and dried saw wood is used as a raw material in manufacturing, moulding, and carpentry. Eucalyptus makes excellent charcoal. The species› leaf extracts can be marketed as a bio-pesticide due to their pesticide-like qualities. Nectar and pollen are abundant in eucalyptus species (Kaur and Monga, 2021). Various uses of *Gliricidia sepium*, including green manure, live fences, firewood fodder, coffee shade, and rat poison, are employed in many tropical and subtropical nations (Nitis *et al.*, 1991). The wood from *Gmelina arborea* is an excellent raw material for many different products, such as pulp, paper, medium-density fibreboard, laminated veneer lumber, wood cement composites, particle board, furniture, furniture interiors, plywood interiors, jointed finger lumber, door panelling, pallets, pencils, and matchsticks. Additionally, it has fair stability, light colour, and homogeneity. Additionally, it is utilised to create high-quality toys and photo frames (KFRI, 2002). *Grewia optiva* branches are trimmed during the winter and submerged in water for a month. The fibre is removed by beating the soaked branches and ropes are made from the extracted fibre. The leaves are used for feeding livestock (Bisht *et al.*, 2006). *Hardwickia binata* produces incredibly dense, hefty, and resilient wood. The wood is mostly used to build bridges, homes, houses, carts, wooden wheels, railway sleepers, beams, and mine supports. Because of their 9-11% crude protein content, leaves are utilised as cattle feed. The bark produces a reddish-brown fibre that can be used for ropes (Korwar, 1994). Jatropha oil from the *Jatropha curcas* tree is a viable and environmentally friendly alternative to traditional fuels like kerosene and diesel. By mixing sodium hydroxide (caustic soda) solution with jatropha oil, jatropha soap can be produced. As seed cake residue, primarily in the form of protein and carbs about 50% of the original seed weight is still present. It can be utilised as fuel fertiliser, or animal feed if it has been detoxified if non-toxic forms are used (Singh *et al.*, 2007). *Leucaena leucocephala* is known as a «miracle tree» worldwide due to its many beneficial uses. For ruminants, the leaves are incredibly nourishing. In addition to being used as paper pulp, the bark can also be made into miniature furniture and parquet flooring and poles can be used as posts, frames, and props for a variety of climbing plants (Brewbaker *et al.*, 1985). Malabar Neem or *Malaivembu* (*Melia dubia)* wood is also used for catamarans, musical instruments, packing cases, ceiling boards, building materials, splints, pencils, matchboxes, agricultural equipment, and tea boxes (Warrier, 2014). Because of the species' versatility, it has a ready and secure market. Additionally, the species is quite adaptable. The plant's fruit is acrid and regarded as anthelmintic. A promising biodiesel tree species is *Pongamia pinnata* (Kesari and Rangan, 2010). Despite not being a frequent practice, the leaves can be utilised as goat and calf fodder. The press cake (seed residue) left over after oil extraction is bitter and unsuitable for use as the only animal feed. Sturdy wood is used to construct combs, tool handles, cartwheels, posts, agricultural equipment, and cabinets. The dying procedure makes use of wood ash. The oil from the seeds is used as a treatment while tanning leather. Seed-derived non-edible oil has been used as a fuel, lubricant, and soap component. Additionally, it is applied as a liniment to rheumatic and cutaneous conditions (Kumar *et al.*, 2015). To supply the match industry in India, *Populus deltoides* was first introduced. Later, plywood made from its wood started to be produced on a large scale. Additionally, it is utilised in packaging fruit and other food products. It is also an excellent source of fibre for various paper grades, including newspaper, packing paper, and fine paper. *Prosopis cineraria* pods are eaten as a vegetable in several regions of India's Thar Desert, both in dried and green forms. The tree provides food during the abnormally dry summer when most trees are leafless. Its pods, which contain a dry, sweet pulp, are also a rich source of feed. Boat frames, homes, poles, and tool handles are all made of its wood (Singh and Bishnoi, 2014). In India, *Salix alba* is widely utilised to make cricket bats. Fruit boxes, particularly apple boxes, are made in Kashmir Valley using wood. In addition, it's utilized to create furnishings, baskets, matchbox splints, agricultural tools, building supplies, sweat lodges, and other items (Mir and Khan, 2008). In addition, it works well for making paper pulp and charcoal. Tender twigs are lopped to feed sheep and goats. *Tectona grandis* wood is renowned worldwide for its durability, tenacity, dimensional stability, working quality, and non-corrosiveness when in contact with metal. It is sometimes referred to as the King of Timbers due to these exceptional qualities (Tewari, 1992). It is mostly employed in manufacturing furniture, cabinets, and different types of shipbuilding, plywood, panelling, piles, construction, poles, and other things. The plywood industry uses increasing amounts of teak to produce high-grade commercial and tea-chest plywood. The seeds have a 44.5% oil content, and the oil is used to make soap. Platters made of teak leaves are popular. Local craftspeople utilize *Terminalia arjuna* for handles on tools, water troughs, boats, carts, and other items (Behera *et al.*, 2015). Construction materials and the plywood industry can both use it. One of the main foods for the tropical *Tasar* silkworm is *Arjuna* leaves. On certain religious festivals, Lord Vishnu and Lord Ganesha receive offerings of the tree's leaves and blooms. The products obtained from different forest plants encourage and add additional income with fewer inputs by the farmers and also motivate the layman to set up an enterprise and obtain benefits under IFS.

Table 1: Promising Agroforestry Plants under Integrated Farming System

Botanical Name	Common Name	Family	Entrepreneurial Opportunities
Acacia mangium	Brown salwood	Fabaceae	Paper Industry, Furniture
Acacia nilotica	Babul	Mimosoideae	Fodder, Gum, Furniture, Paper Industry
Ailanthus excelsa	Maharukh	Simarubiaceae	Paper Industry, Packing Industry, Construction, Fodder for Goats
Alnus nepalensis	Alder	Betulaceae	Fodder
Anthocephalus cadamba	Kadamb	Rubiaceae	Wrapping and printing paper, Fodder, Light construction, match box and pencil industry
Azadirachta indica	Neem	Meliaceae	Lubricant and soap industry, Fodder, Medicinal
Albizia lebbeck	Indian Siris, Kokko, East Indian walnut	Mimosaceae	Resin, Fodder, Honeybee industry and scent industry
Bambusa vulgaris	Bamboo	Poaceae	Handicrafts, furniture, paper industry, agricultural operations
Casuarina equisetifolia	Beef wood	Casuarinaceae	Duplex paper and wrapping paper
Ceiba pentandra	Kapok tree	Bombacaceae	Medicinal, matchbox Industry, Packing industry
Dalbergia sissoo	East India Rose wood	Papilionaceae	Furniture Industry, Fodder
Eucalyptus tereticornis	Eucalyptus	Myrataceae	Carpentry Industry, Bio- pesticide
Gliricidia sepium	Gliricidia	Fabaceae	Fodder, Manure
Gmelina Arborea	Gamhar, White teak	Verbenaceae	Furniture Industry, Carpentry industry, paper industry, match box and pencil industry
Grewia optiva	Bhimal	Tiliaceae	Fibre Industry, Fodder
Hardwickia binata	Indian Black wood	Fabaceae	Furniture Industry, Fibre Industry,
Jatropha curcas	Physic nut	Euphorbiaceae	Fuel and Lubricant Industry
Leucaena leucocephala	Subabul, Ipilipil, Lead tree	Fabaceae	Furniture Industry, Paper Industry
Melia dubia	Malabar Neem	Meliaceae	Packing industry, Match box Industry, Pencil Industry
Pongamia pinnata / Millettia pinnata	Karanj	Fabaceae	Bio diesel Industry
Populus deltoides	Eastern Cotton wood	Salicaceae	Matchbox Industry, Plywood Industry
Prosopis cineraria	Khejadi / Khejri	Fabaceae	Vegetable, Fodder
Salix alba	Willow	Salicaceae	Packing Industry, Furniture Industry
Tectona grandis	Teak	Verbenaceae	Plywood Industry, Furniture Industry
Terminalia arjuna	Arjun	Combretaceae	Plywood Industry, Construction Industry

Tectona grandis *Terminalia arjuna*

Azadirachta indica *Bambusa vulgaris*

Fig. 2: Agroforestry Plants used in Integrated Farming System

Valuable Suggestions Regarding the Incorporation of Agroforestry in an Integrated Farming System (IFS) for Income and Employment Generation

Agroforestry can be included in an integrated farming system (IFS) to provide several advantages, including creating income and jobs and recycling farm by-products. When integrating agroforestry in IFS, keep the following helpful advice in mind:

1. Farm Planning and Design

- To identify appropriate tree and crop species for agroforestry, do a detailed examination of the farm's features, including soil type, climate topography, and water availability.
- Plan the farm's structure to maximize resource use and productivity, considering elements like tree spacing, crop arrangement, and livestock integration.

2. Species Selection

- Select tree species that offer many advantages, such as wood, fruits, nuts, fodder, or therapeutic goods. Consider the local market demand and the species' appropriateness for the agro-climatic conditions.
- Choose locally adapted or native tree species since they are more likely to thrive and need less maintenance.

3. Crop and Livestock Integration

- Choose crops and livestock that can coexist with and benefit from the various tree species. For instance, livestock can graze on crop wastes and shade-tolerant crops grown beneath the forest canopy.
- Use livestock manure as a beneficial organic fertilizer source for trees and crops to improve nutrient cycling and decrease the requirement for outside inputs.

4. Soil and Water Management

- Use soil conservation techniques to prevent erosion and preserve soil fertility, such as contour ploughing, mulching, and cover crops.
- Improve soil water retention using agroforestry systems, especially in drought or water scarcity areas. Windbreaks made of trees can lower evapotranspiration and shield crops from excessive moisture loss.

5. Market-Oriented Approach

- Find markets for agroforestry products, such as wood, fruits, nuts medicinal plants, and value-added goods. To increase earning potential take into account both local and regional markets.
- Form alliances with regional processors, cooperatives, or corporations to ensure a constant market for agroforestry products.

6. Training and Capacity Building

- Train farmers in agroforestry practices, such as tree management intercropping strategies, and sustainable farming methods.
- Encourage farmer collaboration and knowledge exchange through workshops, field trips, or farmer groups to improve their agroforestry knowledge and skills.

7. Research and Monitoring

- Encourage universities, agricultural groups, and research institutions to carry out studies on agroforestry systems with an emphasis on output revenue, and environmental effects.
- Over time, track and assess how well the agroforestry system works, and make necessary changes to management techniques to improve results.

To successfully integrate agroforestry in IFS, remember that careful planning continuing management, and a long-term perspective are all necessary. It is best to get advice from local agricultural specialists or extension agencies who know the market dynamics and your particular agro-climatic conditions.

Conclusion

The practice of integrated farming, which combines the ecological upkeep of a diverse and healthy environment with the economic needs of agriculture helps to maintain a steady supply of healthful, reasonably priced food. Since it is a dynamic idea, it cannot be prescriptive and must be open to change and technological advancements in order to be applicable on any farm, in any nation. Above all, integrated farming is an effective strategy for the future of agriculture that will help all members of society, not just those who engage in it. By integrating economic, ecological, and social goals to the crucial activity of agricultural food production, integrated farming significantly contributes to sustainable development. Crop diversity, synergy, and reciprocal benefits between enterprising components with profitability and sustainability could be fostered by integrating crop livestock agroforestry systems.

References

Ajijur, S., Jette, R., Jacobsen, B., Robert, J., James, H., and Sunderland, T. (2017). Finding alternatives to swidden agriculture: does agroforestry improve livelihood options and reduce pressure on existing forests? Agroforestry Systems, 91(1): 185-199.

Barbieri, C., and Valdivia, C. (2010). Recreational multi-functionality and its implications for agroforestry diffusion. Agroforestry Systems, 79: 5-18.

Behera, L., Nayak, M.R., Patel, D., Mehta, A., Sinha, S.K. and Gunaga, R. (2015). Agroforestry practices for physiological amelioration of salt affected soils. Journal of Plant Science Physiology, 1(1): 13-18.

Bijauliya, R.K., Jain, S.K., Alok, S., Dixit, V.K, Singh, D. and Singh, M. (2017). Dalbergia sissoo Linn. An overview morphology, phytochemistry and pharmacology. International Journal of Pharmaceutical Sciences and Research, 8(4): 1522-1533.

Bisht, J.K., Srivastava, A.K. and Gupta, H.S. (2006). Management of Bhimal (Grewia optiva) planted on terraced field bunds in North Western Himalayas. Indian Journal of Agricultural Sciences, 76(7): 438-440.

Brewbaker, J.L., Hegde, N., Hutton, E.M., Jones, R.J., Lowry, J.B., Moog, F. and Van Den Beldt, R. (1985). Leucaena – Forage Production and Use, NFTA, Hawaii. pp. 1-39.

Darus, H.J.A. (1993). Vegetative Propagation. In: Acacia mangium – Growing and Utilization. MPTS Monograph Series No. 3. Winrock International and FAO, Bangkok, Thailand, pp. 59-74.

Garrity, D.P. (2004). Agroforestry and the achievement of the Millennium Development Goals. Agroforestry Systems, 61: 5-17.

Gill, M.S., Singh, J.P. and Gangwar, K.S. (2009). Integrated Farming System and Agriculture Sustainability. Indian Journal of Agronomy, 54: 128-139.

Hossain, Md., Islam, Md. and Numan, S. (2016). Multipurpose Uses of Bamboo Plants: A Review. International Research Journal of Biological Sciences, 4: 57-60.

Jayanthi, C., Baluswamy, M., Chinnusamy, C. and Mythily, S. (2003). Integrated nutrient supply system of linked components in lowland integrated farming system. Indian Journal Agronomy, 48: 241-246.

Kaur, A. and Monga, R. (2021). Eucalyptus Trees Plantation: A Review on Suitability and their Beneficial Role. IJBSM, 12(1): 16-25.

Kesari, V. and Rangan, L. (2010). Development of Pongamia pinnata as an Alternative Biofuel Crop – Current Status and Scope of Plantations in India. Journal of Crop Science and Biotechnology, 13(3): 127-137.

KFRI (Kerala Forest Research Institute). (2002). Plantation Technology for Nine Indigenous Tree Species of Kerala. KFRI Research Report 231.

Korwar, G.R. (1994). Hardwickia binata – A Promising MPTS for Agroforestry in Dryland Areas. In: Multipurpose Tree Species for Agroforestry in India. Hegde, N.G. and J. N. Daniel (Eds). BAIF Development Research Foundation 'Kamdhenu', Senapati Bapat Road, Pune, India.

Krisnawati, H., Kallio, M. and Kanninen, M. (2011). Acacia mangium Wild: ecology, silviculture and productivity. CIFOR, Bogor, Indonesia, pp. 1-15.

Kumar, A., Suvera, H., Thakur, N.S. and Jha, S.K. (2015). Herbage and essential oil yield of Ocimum spp. intercropped under Pongamia pinnata based silvi-medicinal systems in Gujarat, India. The Bioscan, 10(1): 81-85.

Mir, A.H. and Khan, M.A. (2008). Projected benefit-cost analysis of agri-silvicultural system: vegetable crops intercropping with Salix alba (Willow). Journal of Forestry Research, 19(4):335-339.

Nicodemus, A. (2014). Cultivation techniques for Casuarina. In: Transfer of Tree Cultivation Technologies to Krishi Vigyan Kendras (KVKs) of Tamil Nadu and Puducherry. Institute of Forest Genetics and Tree Breeding, Coimbatore, India, pp. 6-11.

Nitis, I.M., Lana, K., Suarna, M., Puger, A.W., Sukanten, W. and Putra, S. (1991). Farm level seed production of a top performing Gliricidia sepium in dry-land farming area in Bali, Indonesia. Technical Report prepared for FAO, Rome. pp. 1-36.

Orwa, C., Mutua, A., Kindt, R., Jamnadass, R. and Anthony, S. (2009). Agroforestry tree Database: a tree reference and selection guide version 4.0 [http://www.worldagroforestry.org/sites/treedbs/treedatabases.asp]

Pankaj, S., Lokeshwar, T., Mukesh, B., and Vishnu, B. (2011). Review on neem (Azadirachta indica): Thousand problems one solution. International Research Journal of Pharmacy, 2. 97-102.

Rajasugunasekar, D. (2014). Cultivation techniques of Ailanthus excelsa. In: Transfer of Tree Cultivation Technologies to Krishi Vigyan Kendras (KVKs) of Tamil Nadu and Puducherry. Institute of Forest Genetics and Tree Breeding, Coimbatore, India, pp. 19-22.

Reddy, N.R. (2021). Medicinal and Aromatic Plants for Enterprise development. In: Webinar on Cultivation of Medicinal Plants in the context of COVID -19: Issues and Challenges. ICAR-Directorate of Medicinal and Aromatic Plants Research, Anand, Gujarat.

Sharma, R., Xu, J. and Sharma, G. (2007). Traditional agroforestry in the eastern Himalayan region: Land management system supporting ecosystem services. Tropical Ecology, 48(2): 189-200.

Shekinah, D. E., Jayanthi, C. and Sankaran, N. (2005). Physical indicators of sustainability – A farming systems approach for the small farmer in the rainfed vertisols of western zone of Tamil Nadu. Journal of Sustainable Agriculture, 25: 43-65.

Singh, J. and Bishnoi, M. (2014). Effect of lopping intensities on fodder and fuel wood yield of Prosopis cineraria in arid zones of Thar. Academia Arena, 6(10): 90-94.

Singh, P. and Agrawal, T. (2018). Albizia lebbeck (L.) Benth: A Short Review. International Journal of Scientific Research in Chemical Sciences, 5(5): 08-11.

Singh, R.A., Kumar, M. and Haider, E. (2007). Synergistic cropping of summer groundnut with Jatropha curcas – A new two-tier cropping system for Uttar Pradesh. Journal of SAT Agricultural Research, 5(1): 1-2.

Subapriya, R. and Nagini, S. (2005). Medicinal properties of neem leaves: A review. Current Medicinal Chemistry – Anti-Cancer Agents. 5(2): 149-156.

Tewari, D.N. (1992). A monograph on teak (Tectona grandis Linn. f.). International Book Distributors, Dehradun, India, pp. 1-479.

Vijayaraghavan, A. (2014). Cultivation Techniques for Neolamarckia cadamba. In: Transfer of Tree Cultivation Technologies to Krishi Vigyan Kendras (KVKs) of Tamil Nadu and Puducherry. Institute of Forest Genetics and Tree Breeding, Coimbatore, India, pp. 28-32.

Viswanath, S., Nair, P.K.R., Kaushik, P. and Prakasam, U. (2000). Acacia nilotica trees in rice fields: A traditional agroforestry system in central India. Agroforestry Systems, 50: 157–177.

Warrier, R.R. (2014). Cultivation techniques for Melia dubia. In: Transfer of Tree Cultivation Technologies to Krishi Vigyan Kendras (KVKs) of Tamil Nadu and Puducherry. Institute of Forest Genetics and Tree Breeding, Coimbatore, India, pp. 16-18.

9

Lac Centric trio-IFS Models to Boost Entrepreneurial Development Among the Tribal in India

Neelkusum Xess[1] and Sudhanand Prasad Lal[2]

[1]ICAR- National Dairy Research Institute, Eastern Regional Station Kalyani, West Bengal

[2]PG Department of Extension Education, Dr. Rajendra Prasad Central Agricultural University Pusa, Samastipur, Bihar

Abstract

As there is a decline in the average landholding of individual farmers in India, it is important to take precautionary steps to improve and sustain farm income for their growth and development. One of the salient features which can be adopted in farming situation is going for integrated farming system. It not only reduces the expansion of land activities but also helps in input cost reduction and interdependent resource utilisation in the farm activity. Lac is a natural secretion of the lac insect (Kerria lacca) and is a NTFP (Non- Timber Forest Product) and agro based activity which produces different types of value-added products like shellac, lac dye, dewaxed lac, button lac, shellac wax etc. India is one of the largest exporters of lac in the world. Its value-added products have a heavy demand in the global trade market. Integrated farming systems can be used as one of the major components. The earlier studies have shown significant profits from allied activities. It has been reported that 56.25% of farmers are engaged in lac cultivation alongside agriculture with the majority belonging to Scheduled Tribes. Lac can be integrated with different components such as horticultural crops, apiary, sericulture, goatry, poultry, livestock, and mushroom cultivation to enhance productivity and support farmers in sustaining income. It is an eco-friendly activity which has caught an eye for progressive farmers and entrepreneurs. States like Jharkhand, Chhattisgarh, West Bengal, Maharashtra and Madhya Pradesh are the major lac producers as well as processors in our country. It is a highly remunerative crop which gives employment in the off season to farmers. So, it can be combined with other allied activities in farming which work in favour of income enhancement as well as help in the betterment of present faming scenario in the nation.

***Keywords*:** Agriculture, IFS, Lac cultivation, Non-timber forest product, Income enhancement.

Introduction

India is an agriculture dominated country. According to the report of the Situation Assessment Survey of Agricultural Households by the NSO, 54.6% of the total workforce is engaged in agriculture and allied sectors which encompasses for 17.8% of India's total GVA for the year 2019-20. The number of operational holdings in the country is estimated at 14.64 crores (Agriculture Census 2016).

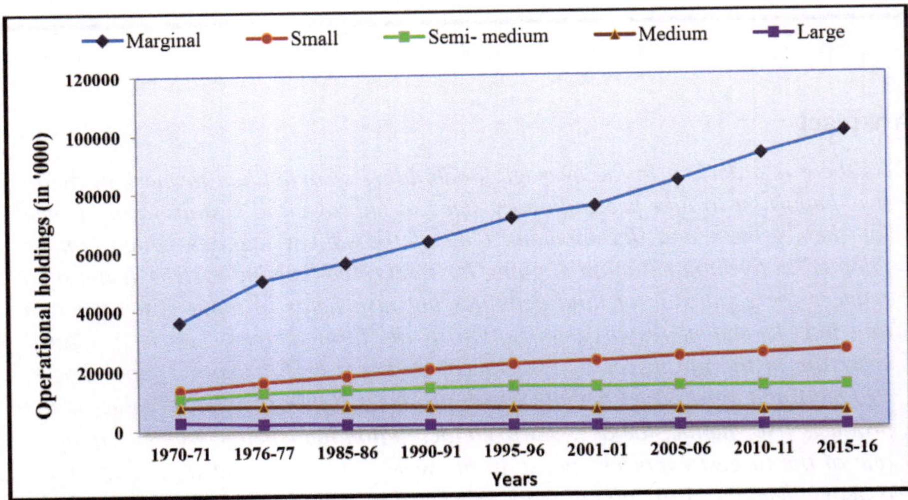

Fig. 1: Average land holding of farmers [*Source*: GOI, 2021]

As we can see from Figure 1 the operational landholding of Indian farmers is increasing day by day. The average size of operational holdings was 1.15 hectares in 2010-11 which has declined to 1.08 hectares in the year 2015-16 (Agriculture Statistics at a Glance, 2021). This indicates that land fragmentation is becoming more prevalent. If there is more fragmentation of land it will lead to a reduction in the per capita holding of an individual farmer. This situation will significantly affect agriculture stakeholders, especially small and marginal farmers by reducing their market access, information accessibility, transport costs, machinery usage, access to extension advisory services, negotiation power, selling and purchasing power, and import and export of products. (Agriculture Census, 2015-16).

As there is a gradual reduction in land holdings, an increase in demand for food supplies and the need to sustain and improve the income of farmers, integrated enterprises need to be incorporated into the current farming pattern. Commodities such as fisheries, poultry, duckery, sericulture, apiculture, livestock, horticultural crops, or lac culture can be added. The farming system approach may play a vital role in enhancing the farm profitability, productivity and sustainability for small and marginal farmers. Mahapatra (1994) outlined that soil, water, crops, livestock, labour, capital, energy and other resources are taken into account in the farming system at the centre for managing agricultural and related activities. The farming system involves complex interactions among various interdependent components. The need for a farming system approach is to develop farms, increase farm production and sustain the livelihoods of farm families by increasing their income.

Present Scenario of Lac and its Importance

Lac is a fauna-based natural resin. It is one of the most popular natural resin secreted generally by the Indian lac insect, *Kerria lacca*, that lives on the tender branches of particular host trees *such as Flemingia semialata,* a preferred and promoted plant for scientific lac cultivation by IINRG, kusum (*Schleichera oleosa*) which is the most acquired tree by the majority of farmers, palas (*Butea monosperma*), ber (*Ziziphus mauritiana*), etc. On the basis of the choice of lac insect, there are two main strains of lac insects, Rangeeni and Kusumi. The Kusumi strain is grown on kusum trees. The life cycle of lac insects takes around 6 months, therefore, 2 crops a year can be grown. In Kusumi strain, Jethwi (Jan/Feb) and Aghani (June/July) crops are taken. In Rangeeni, Katki (June/July) and Baisakhi (Oct/Nov) crops are taken. The lac of Rangeeni crops is harvested when immature. Baisakhi and Aghani of Rangeeni and Kusumi strains are the chief crops which contribute around 90 percent of lac production. On the other hand, the Kusumi crop lac is seen as a better resin, as it is light in colour and fetches a good price.

Resin, dye and wax are the three important products of raw lac. According to the Forest Survey of India (2007) there are about 49.2 crores of palas, 4.5 crores of ber and 3.7 crores of kusum spread all over the country in various agroforestry systems which account for 3% of the total number available in India. Lac cultivation is one of the vital sources of revenue for the livelihood of the people in the woodland and sub-forest dwellers of Jharkhand, Chhattisgarh, and nearby states. It is a labour-intensive process and employs both genders living in the woodland and sub-woodland areas of these regions. Our nation is the largest producer of lac. However, its production in India during the 2010s the efforts in terms of research, policy, and growth concerning the upliftment of this sector were limited. It has been observed that less than 10% of lac host trees are being used for lac production (Lac, Plant Resins and Gums Statistics at a Glance, 2019). It is understood that some barriers are hindering the growth of the sector.

Table 1: Lac Production and Share of States in India

Name of the States	Name of the Lac/Crop strain						Weight (tonnes)	Percentage Share
	Baisakhi	Katki	Rangeeni	Jethwi	Aghani	Kusumi		
Jharkhand	634	286	920	4640	4991	9631	10551	56.92
Chhattisgarh	760	505	1265	550	1635	2185	3450	18.61
Madhya Pradesh	1222	826	2048	42	46	88	2136	11.52
West Bengal	69	610	679	80	203	283	962	5.19
Maharashtra	295	405	700	7	15	22	722	3.90
Odisha	12.5	9	21.5	189	436	625	646.5	3.49
Total	2992.5	2641	5633.5	5508	7326	12834	18467.5	99.63
Overall	3029.5	2661	5690.5	5513	7333	12846	18536.5	(100.00)

These states contribute to 99.63% share of total production of lac in India. The states of Assam, Meghalaya, Uttar Pradesh and Andhra Pradesh contribute less than 0.5% (precisely 0.37%) all together.

Table 2: Lac Production and Ranks of Different Districts in India

District	2017-18	Rank	2018-19	Rank
Ranchi	2450	1	3170	1
Simdega	1855	2	2330	2
Khunti	1115	3	1470	3
Gumla	995	4	1430	4
Korba	630	9	885	5
Kanker	410	10	855	6
West Singhbhum	670	7	845	7
Seoni	864	5	820	8
Balaghat	746	6	697	9
Gondia	660	8	570	10

Jharkhand is the top producer state of lac, holding a 56.92% share among all other producing states (Table 1). In 2017-18 and 2018-19, Ranchi has been holding first rank in the production lac followed by other districts like Simdega, Khunti and Gumla, respectively. Most of the farmers are small and marginal who are involved in lac cultivation. This provides them with income and employment throughout the year. Lac cultivation can be done all the seasons so it can be adopted as an extra enterprise. Rice cropping is prevalent in eastern India but for lac cultivation we don't need extra land. During last few years although there is an increase in the total production of lac but fewer people are adopting this cultivation. From 2010-11 to 2018-19, lac exports on average per year were 6923 tonnes worth at ₹32012 lakhs. According to one of the reports of IINRG, there was a positive trend in the annual compound growth rate of exported quality lac (3.40%). But there has been a decline in growth rate in the value terms (5.22%) (Figure 2).

The declining rate in terms of value may be due to various reasons such as reduction in quality, cultivation practices, low yield and income and emergence of artificial synthetic dyes. These problems can be reduced by applying various technologies and one of them the integration of farming system. A single enterprise is not enough to sustain the farm family, but the farming system approach is a promising attribute to address the problems and solve the issues of sustainable economic growth in the Indian farming community. The integration of crops with animal husbandry, poultry, duckery, and piggery is common in agricultural practices. However, other enterprises like sericulture, apiculture or lac culture can bring monetary benefits as well as employment throughout the year without the demand for extra land or space.

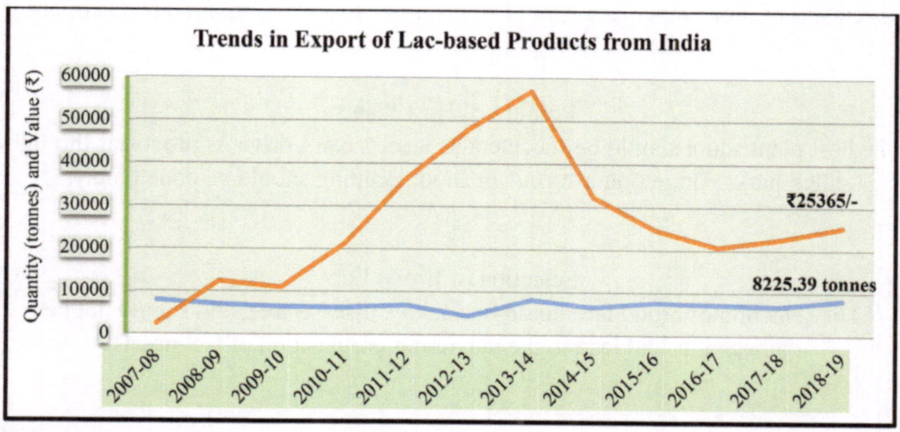

Fig. 2: Trends in export of lac-based products from India

Existing lac-based cropping systems and their importance in term of productivity profitability and nutrition

There are two methods – Local and Modern cultivation – through which lac is being cultivated in India.

1. **Local Cultivation:** The cultivation practices adopted by the lac growers in India are analogous, apart from slight variations. Some shoots are left on the same tree after taking recurring partial lac crops so that lac can be inoculated on other branches. The procedure is repeated at the time of harvesting. This method does not provide a continuous supply of brood lac for inoculations and also cannot generate an optimum yield of lac.
2. **Modern Cultivation proposed by the Indian Institute of Natural Resins and Gums, Ranchi (IINRG):** IINRG, Ranchi is an institute under the Indian Council of Agricultural Research which was ranked 62nd in the Ranking of institutes in the year 2019-20 and 2020-21. They have developed a scientific method of cultivating lac which will be helpful to lac growers. The following steps should be followed during lac cultivation using this method–

Selection of a Suitable Host Plant

Firstly, we should select a suitable host plant according to strain for lac cultivation.

↓

Selection of Suitable Site for Lac Cultivation

It is followed by a selection of a suitable location. It is that it can be grown only in open spaces. For proper care should be taken for free air circulation and sunlight.

↓

Pruning of Host Plants

The host plant shoot should be succulent as insects can't drive its proboscis through thick bark. For getting a succulent shoot, pruning should be done timely.

↓

Selection of Brood Lac

The selection of brood lac should be healthy, disease-free, and should not be damaged. It will lead to the maximum propagation of lac insects.

↓

Inoculation of Brood Lac in Host Trees

It should be done at the proper time. Bundle of brood lac should keep inside 60 mesh nylon net bag.

↓

Inoculation of Brood Lac in Host Trees

It should be done at the proper time. Bundle of brood lac should keep inside 60 mesh nylon net bag.

↓

Phunki Removal
Phunki removal is another technique that should be performed. After the emergence of insects, the brood lac is removed after 21 days.

↓

Harvesting of Raw Lac
It is the next important operation which should perform after the appearance of yellow spots, which appears about a week before the emergence of the larva.

↓

Scrapping of Lac from Stick
After harvesting the encrusted lac is scrapped through a knife and it can be stored and further processing can be done.

↓

Processing of Lac
Lac can be processed with the help of various instruments and numerous forms of lac are developed which are further useful in industry.

Fig. 3: Scientific Method of Cultivation Recommended by IINRG, Ranchi

Application of Lac

It is non-hazardous and has edible properties. The demand for lac is increasing day by day. It is used in the car industry, electrical industry, paint industry, wood industry, pharmaceutical industry, and other industries for decorative uses among others. Among the lac-cultivating states, Jharkhand is the highest producer contributing about 56% of the national lac production.

Fig. 4: (a) Brood lac used for inoculation; **(b)** Inoculation with brood lac in host

Recent studies from the Indian Institute of Natural Resins and Gums show that 28% of the total agricultural income comes from lac production. However, the farmers involved are the poorest among the state. Despite a large number of host trees, only a few are used for lac cultivation (Kumari and Jaiswal, 2019).

Table 3: Lac Production with Allied Enterprises

S. No.	Category	*f* (n=80)	% share
1.	Lac production	1	1.25
2.	Lac production + Agriculture	45	56.25
3.	Lac production + Agriculture + Labour	13	16.25
4.	Lac production + Agriculture + Animal rearing	7	8.75
5.	Lac production + Agriculture + Business	9	11.25
6.	Lac production + Agriculture + Service	3	3.25
7.	Lac production + Other enterprises	2	2.50

According to a study conducted in 2020, a total of 80 respondents from two blocks of Ranchi district were selected. The data revealed that 56.25% of respondents are engaged in lac cultivation and agriculture, which is the highest in the sample size. Meanwhile, 16.25% of respondents belong to the 3rd category which includes lac cultivation, farming, and labour work; and 11.23% belong to lac production, agriculture, and small business. About 8.75% are engaged in animal rearing along with lac cultivation and only 3.75% of respondents have jobs or services along with the cultivation of lac. Almost 2.50% of respondents are fully dependent only on lac cultivation and 1.25% of respondents have other services. It shows that the majority of farmers are involved in lac production and crop production. This shows that there is an immense need to make people aware of taking up integrated enterprises in order to have a sustainable livelihood. In the same study, it was also recorded that the majority of the respondents (56.25%) belonged to the Scheduled Tribe, followed by the Unreserved, Other Backward Class and Scheduled Caste respectively.

Table 4: Important Constraints related to Lac Cultivation (n=80)

Sl. No.	Constraints	Mean	Rank
1	Ecological constraints	1.43	1
2	Cultural constraints	3.31	2
3	Technical constraints	3.99	3
4	Extension constraints	4.54	4
5	Financial constraints	4.99	5
6	Technological constraints	5.24	6
7	Marketing constraints	6.45	7
8	Input constraints	7.95	8
9	Labour constraints	8.16	9
10	Social constraints	9.10	10

There was a total of different constraints which were grouped in 10 categories like technical, input, marketing, financial, ecological, technological, social, labour, cultural, and extension constraints. The top constraint was ecological constraints with a mean value of 1.43 was given rank I followed by cultural constraints with a mean value of 3.31, technical constraints with a mean value of 3.99, extension constraints with a mean value of 4.54, and financial constraints with a mean value of 4.99. Meanwhile, technological constraints, marketing constraints and input constraints were given rank VI, VII, and VIII with mean values of 5.24, 6.45, and 7.95 respectively. Labour constraints were given rank IX with a mean value of 8.16. The last constraint was social constraints with the highest mean value of 9.1 with rank given X.

Improvement in the Existing Lac-based Farming System

It can be done on the basis of selection of suitable enterprises for having a sustained lac-based cropping system.

1) Lac Cultivation with Integration of Vegetable Crops: A survey was conducted by Khan (2018) by taking 57 BPL families in Ranchi district. Most of the families were dependent on agricultural crops and animal rearing along with lac cultivation. A total of 9 types of farming systems was taken out of which B:C ratio was calculated for 5 types based on the number of farmers. Out of the 9 systems, only 5 were selected on the basis of relevance.

Table 6: Economics of Different Farming System [Source: Khan, 2018]

Farming System	Type of Farming System	Total Investment (₹)	Total Return (₹)	Net Income (₹)	B:C Ratio
FS-1	Lac + Paddy + Tomato + Bottle gourd + Goatry + Poultry	109748	346550	236802	3.15
FS-2	Lac + Paddy + Tomato + Goatry + Poultry	86744	282550	195806	3.25
FS-3	Lac + Paddy + Tomato + Cabbage + Goatry + Poultry	114688	382550	267862	3.33
FS-4	Paddy + Tomato + Goatry + Poultry	72119	208100	135981	2.88
FS-5	Lac + Paddy + Tomato + Bottle gourd	91073	308450	217377	3.38

Table 7: Lac Cultivation with Integration of Horticultural Crops [Source: Kumar, 2013]

Cropping System	No. of plants	Ratio Of plants	B:C Ratio in 1st Year
Flemingia semialata + Papaya + Mango	7200 – *Flemingia semialata* 1200 – Papaya 600 – Mango	*Flemingia semialata*: Papaya: Mango 6:2:1	3.65
Flemingia semialata + Papaya + Mango + Kusum	7200 – *Flemingia semialata* 900 – Papaya 450 – Mango 450 – Kusum	*Flemingia semialata*: Papaya: Mango: Kusum 16: 2: 1	3.26.
Flemingia semialata + Papaya + Mango + Kusum	4500 – *Flemingia semialata* 433 – Kusum 433 – Papaya 433 – Mango	Kusum: Papaya: Mango 1:1:1	2.68

A survey conducted by Kumar (2013) shows that integration with horticultural crops has resulted in better returns. From the first year onwards, the B:C ratio seems to be around 3 and it is expected to increase in the upcoming years.

Strategies for Improvement through Lac-based Integrated Farming System

IFS Model-1: Lac cultivation + Goatry + Mushroom

The IFS model assumed in Figure 6 depicts the integration of different enterprises like goatry, mushroom cultivation and agricultural crops with lac cultivation. This suggested model can enhance the farm economics as the goat milk is high in nutrients and it has a high demand in the market. The meat from goats is also supposed to bring some income. The mushroom cultivation can also be done together as it doesn't need large area. It can be adjusted in a dark small room with the regulation of appropriate temperature. It can also bring some monetary benefits for farmers.

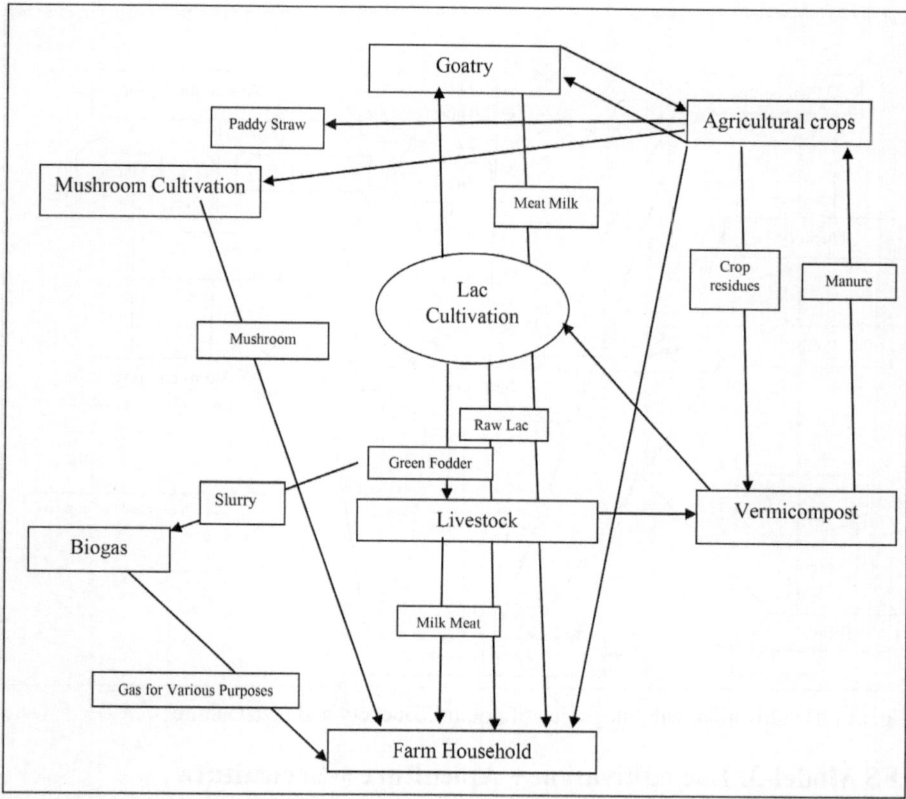

Fig. 5: Lac cultivation with Integration of Goatry and Mushroom Cultivation

IFS Model-2: Lac cultivation + Poultry + Duckery + Fishery

The suggested model in Figure 7 integrates poultry, fishery, and fish culture with lac cultivation. This model can be beneficial as these three components are rich in protein and their management can be simple. All the waste resources can be easily applied as input for the others. This model can be used for nutrition enhancement and can provide extra income by selling the products.

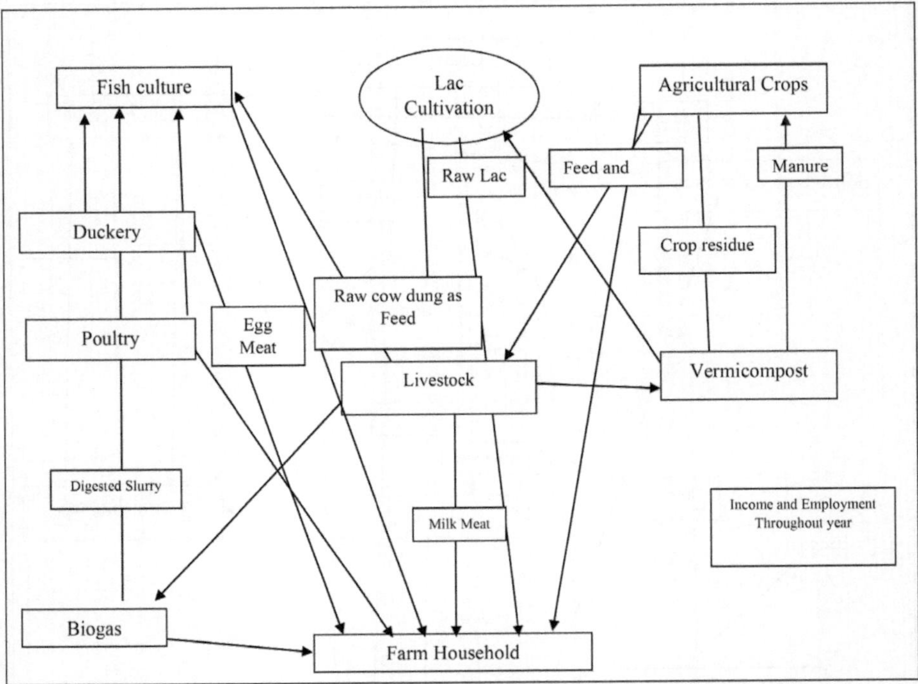

Fig. 6: Lac cultivation with Integration of Poultry, Duckery and Fish Culture

IFS Model-3: Lac cultivation + Apiculture + Sericulture

This proposed model involves integrating with apiculture and sericulture. Apiculture is beneficial for pollination, which can help lac host plants grow. During pruning, branches are trimmed and the lac insect generally secretes resin in tender branches. Therefore, bees in apiculture could serve as pollinators for lac host plants.

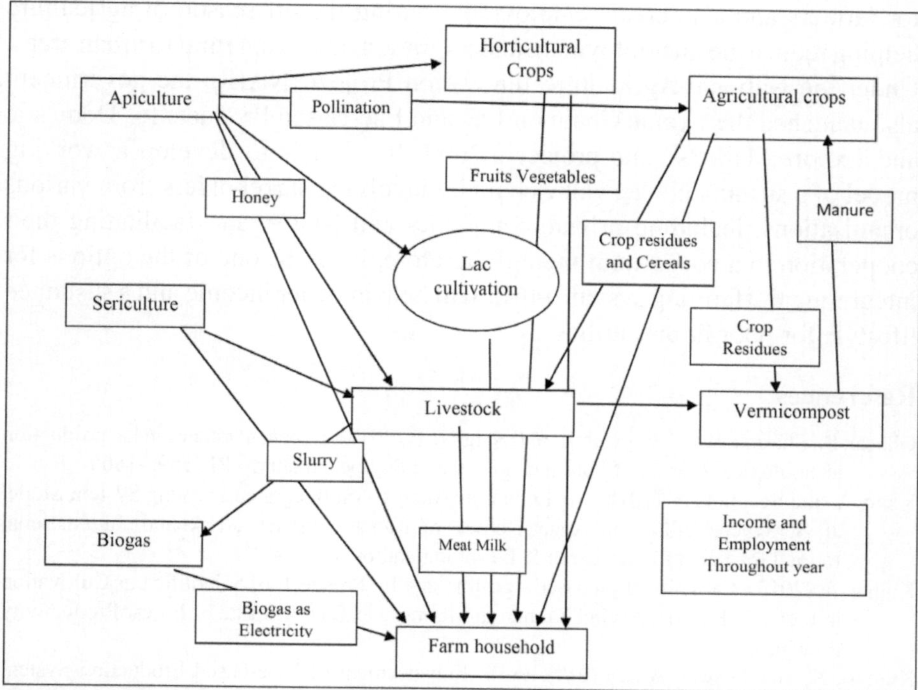

Fig. 7: Integration of Lac Cultivation with Apiculture and Sericulture

Conclusion

Lac culture involves the cultivation of raw lac which is processed into different types such as Shellac, Dewaxed lac and Seed lac for its several purposes in industries. It is an agro-based activity which can be undertaken by any kind of farmer as it does not require much land. Lac cultivation can be done on trees which are Kusum, Ber etc. which are already present. It is a very lucrative crop, providing substantial economic returns to farmers and contributing to the country's foreign exchange through exports. There is a huge demand for lac and its value-added products in countries like Bangladesh, the USA, and Pakistan. The compound growth rate in exported quantity was highest for lac among all the natural resins and gums since the year 2012-13. A surge in demand has also been observed for lac processed products. Therefore, it can be adopted as one of the income-producing activities among farmers especially for small and marginal farmers. It can be considered as one of the enterprises which does not need any kind of huge investment or require land. The major focus can be given in creating general awareness about its profitability, cultivating lac using scientific methodology provided by IINRG and integrating with other allied enterprises. This can generate extra income for farmers and also create employment during the off-season of agriculture helping reduce the unemployment and migration rates from rural to urban areas. Under the National Agriculture Innovation Project (NAIP), the government also launched the "Value Chain on Lac and Lac-Based Products for Domestic and Export Markets" sub-project in2009. Its goal is to develop a working model of a sustainable lac value chain by involving stakeholders from various organisations, including private companies and NGOs, and facilitating their cooperation in a consortium mode. Therefore, it can be one of the options for integrating the farming system which will help in better income and a sustained lifestyle for poor farm families.

References

Ghosal, S., Sharma, K.K., Singh, A.K. and Singh B.P. (2018). Recent advances in lac production technologies. Journal of Pharmacognosy and Phytochemistry. SP1: 1598-1600.

Khan, A. and Sengupta A. (2018). An Economic Analysis on Integrated Farming System Model of Burakocha village in Angara block of Ranchi District, Jharkhand. International Journal for Scientific Research & Development. 6(7): 474-477.

Kumar, A. (2013). Lac cultivation based Agroforestry, In: Prospects of Scientific Lac Cultivation in India, 1st Edition, Arvind Kumar and Rameswar Das, Institute of Forest Productivity Ranchi, India.

Kumari, K. and Jaiswal, A.K. (2019). Profit Enhancement of Rice-based Production System through Lac Enterprise. Current Journal of Applied Science and Technology. 38(6): 1-5.

Ministry of Agriculture & Farmers Welfare (2019). Agriculture Census 205-16. Agriculture Census Division, Department of Agriculture, Co-Operation & Farmers Welfare, Government of India.

Ministry of Agriculture & Farmers Welfare (2022). Agricultural Statistics at a Glance 2021 Department of Agriculture & Farmers Welfare, Directorate of Economics & Statistics, GOI, New Delhi.

Ministry of Information and Broadcasting (2022). India State of Forest Report-2020. (RU-49-01-0047-200122). Press Information Bureau. Government of India.

Pal, G., Prasad, N., Ghosal, N., Pandey, S.K. and Kumar, M. (2013). Socio-economic profile of lac growers in Ranchi and Khunti districts of Jharkhand. The Journal of Rural and Agricultural Research. 13(2): 1-5.

Sarkar, P.K., Yadav, V.K., Chakrabarti, A., Kumar, P.R., Shinde, R., Thombare, N., Jha, B.K., Das, B., Dhakar, M.K., Singh, A.K. and Bhatt, B.P. (2020). Lac-based agroforestry models for prosperity of farmers in Jharkhand. Agriculture & Food: e-Newsletter. 2(2): 576-581.

Singh, A.K., Jaiswal, A.K., Bhattacharya, A., Singh, R.K., Monobrullah, Md. and Ghosal, S. (2015). Maximization of profitability through lac production on Flemingia semialata – A Bushy Lac Host. VEGETOS: An International Journal of Plant Research & Biotechnology. 28(2): 219-222.

Singh, B.K., Oraon, P.R., Ranjan, S., Kumar, A. and Suman, K.K. (2019). Survey of existing farming system and agroforestry practices in the Lohardaga District of Jharkhand, India. International Journal of Current Microbiology and Applied Sciences. 8(9): 1625-1633.

Xess, N. (2020). Study on Extent of Adoption of Lac Technology among the Growers of Ranchi District of Jharkhand, M.Sc. Thesis, Dr. Rajendra Prasad Central Agricultural University Pusa, Samastipur, Bihar.

Yogi, R.K., Kumar, N. and Sharma, K.K. (2021). Lac, Plant Resins and Gums Statistics-2019: At a Glance. ICAR-Indian Institute of Natural Resins and Gums, Ranchi (Jharkhand) India. Bulletin (Technical) No. 3/2021, pp. 01-82.

Yogi, R.K., Sharma, K.K. and Ramani, R. (2016). Model Bankable Projects: Lac Cultivation for Livelihood Security. ICAR-Indian Institute of Natural Resins and Gums, Ranchi, Jharkhand. Bulletin (Technical) No.: 17/2016, pp. 1-34

10

Entrepreneurial Opportunities in Sericulture-based Integrated Farming System

Chethan Babu R.T.[1], Praveen B.R.[2], Dileep R.[3], Rakesh Kumar[1] and Birendra Kumar[1]

[1]Agronomy Section, ICAR-National Dairy Research Institute Karnal, Haryana
[2]Scientist B, Tea Research Association (North Bengal Regional R&D Centre), Nagarakata, West Bengal
[3]Department of Soil Science and Agricultural Chemistry, U.A.S. Raichur, Karnataka

Abstract

Sericulture is an agro-based cottage industry that is highly suitable for farmers in rural areas and serves as a subsidiary occupation in growing regions. Sericulture-based integrated farming systems (IFS) involve combining various farm units such as crop production, dairy, vermicomposting and poultry into a single unit. Fish farming has wide scope for profit-oriented entrepreneurship opportunities. Sericulture is a broad term comprising seed production, chawki rearing, host plant cultivation, cocoon production, reeling, crafting, marketing silk products etc and value addition of sericulture-based by-products provides economically viable entrepreneurship with outstanding job opportunities to rural as well as skilled youth in semi urban region and have significant effect on economic and social condition of people. Practicing scientific technology, public and private sector involvement in entrepreneurship development in this make more profitable and sustainable. This chapter will provide a critical overview of venture opportunities for start-ups in the sericulture based integrated farming system.

Keywords: Sericulture, IFS, Entrepreneurship, Mulberry, Cottage industry.

Introduction

India ranks second in global silk production next to China and first in raw silk and silk fabrics consumption, accounting for 70% with a production of 34,923MT in 2021-22. In total silk production traditional crossbreed silkworms contribute 3/4th and 1/4th bivoltine breed (CSB reports). An important strategy for creating jobs and offering a variety of entrepreneurial prospects for rural development is sericulture, which has a solid agricultural foundation. Sericulture is comprised of both agro-based and industry-based activities. Agro-based activities include mulberry production and rearing of silkworms while industrial activities include post-cocoon activities. The different entrepreneurial fields in sericulture include mulberry seedling nursery, production of silkworm eggs (DFLs), chawki rearing units, cocoon production, silk reeling, twisting, silk yarn and fabric dyeing, printing, weaving, fabric finishing and other silk-based handicrafts. Through well-organised marketing, sericulture can be a potential employment generator and income earner in India, especially for rural women.

Sericulture is one of India's economic cottage forest-based industries with high labour requirements in traditional sectors under small-scale markets. Silk is the final economical product of sericulture. It is tailor-made activity for rural poor farmers, entrepreneurs and artisans because of its lower investment. This occupation helps in uplifting rural poor, particularly marginal farmers and other weaker and less fortunate segments by providing year-around income and employment generation. Sericulture is a multi-layered industry that includes moriculture, worm rearing, grainage, reeling, printing and dyeing, fabrication, textile industry, and marketing of all the by-products. Sericulture embodies a wide variety of on-farm and off-farm events, collaborating various field expertise as one to work in the silk industry. Sericulture is a low-investment, high-yielding crop with a short duration that facilitates taking five to six crops per annum. Sericulture produces as many valuable or utilisable by-products and these by-products have utility in different other enterprises. Integration of sericulture with different agriculture allied sectors like crop husbandry, dairying, pisciculture, agro-forestry has substantial potential in agriculture entrepreneurship. Besides income generation it also assists the farming community with effective employment generation for family members. The integrated farming system (IFS) approach alters agricultural operations for higher productivity in the cropping system while also ensuring optimum utilisation of on-farm resources. In this system, the output or waste of one enterprise is utilised as input through better recycling for profitable purposes. Thus sericulture-based IFS provides a wide range of job opportunities and entrepreneurship activities.

Integrated farming systems are biologically defined as "the integration of natural resources and mechanisms into farming activities to attain maximum replacement of external inputs with the aim of sustainable production of high-quality food, fodder and other products through ecologically viable technologies. This approach sustains farm family income, reduces environmental impact, and offers other agricultural benefits (IOBC, 1983). This section provides a glimpse of the business opportunities in agriculture and the allied sectors in the Islands for entrepreneurship development in farming systems.

Agricultural entrepreneurship is a process of innovation that redistributes resources and frequently creates new opportunities by combining uncommon resources and risk-taking abilities. It can produce a wide range of economic benefits in the agricultural sector, including higher agricultural productivity, the formation of new businesses, creating new jobs, the development of rural areas and innovative goods and services. An entrepreneur is someone who organises production, takes risks, and manages the financial insecurity that comes with running a business and also has the desire to create or accomplish something unique. These collective attributes of the entrepreneur are termed as entrepreneurship. Agripreneurship is a dynamic business manager who conducts numerous agri-based operations including physical, financial and human resources to accomplish a particular goal.

Scope for Entrepreneurship in Sericulture-based Integrated Farming System

Entrepreneurship in sericulture-based IFS comprises collective innovative development of moriculture, sericulture, agriculture crop husbandry, animal husbandry, poultry and horticulture etc. in an integrated manner to increase agricultural productivity for sustainable improvement and wealth of farmers involved in their relevant sectors. Considering the fact that integrated implementation of various agri-enterprises in a well-planned strategy with optimum efforts yields better results than any of the enterprises alone. Therefore, sericulture is the only enterprise that can be easily integrated with any of the others and offers tremendous scope and opportunities for farmers to elevate their socio-economic status by generating additional income.

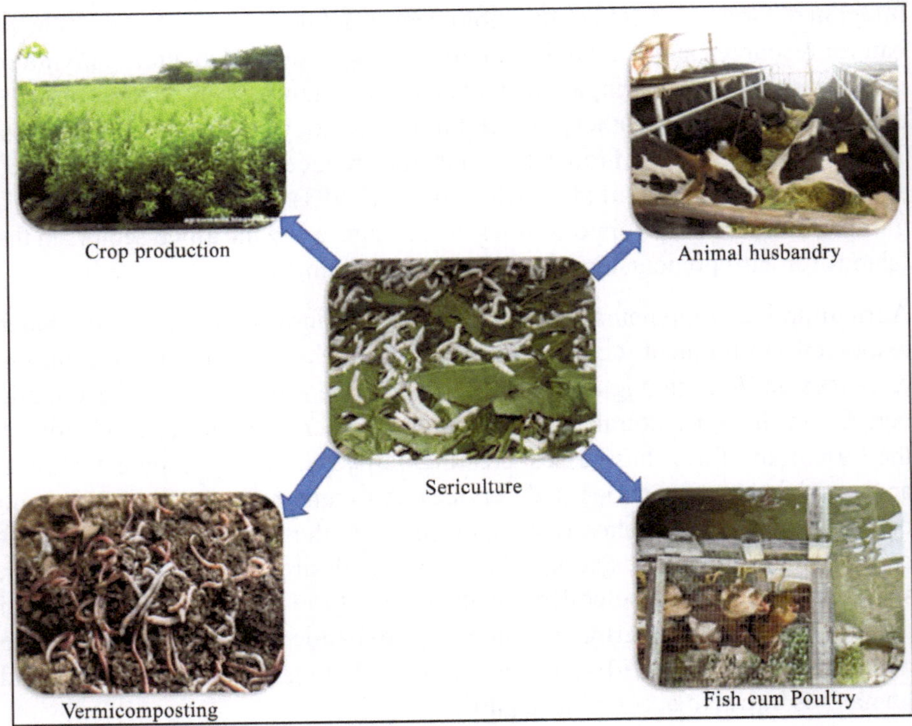

Fig. 1: Sericulture-based integrated farming system model

Dairy Farming: Through a well-planned management strategy various small enterprises like sericulture, moriculture and dairy farming can be integrated which would increase the involvement of labour folk and offer opportunities for farmers to gain additional benefits. Sericulture produces a considerable quantity of residues in the field as well as the rearing house. These residues can be used effectively as feed for small-scale livestock. Thus, farmers can save on the cost of feed and earn extra income from milk and meat along with income from sericulture. In an experiment conducted by Nagaraju and Raghavendra (2016), it has been reported that the adoption of integrated farming, particularly in combination with crops, dairy, sheep and sericulture generated the most beneficial results for farmers in terms of both income and employment generation which was recorded as 322man days per year.

Poultry cum Fish Farming: In India sericulture is viewed as a subsidiary occupation by rural farmers. Moreover, the seasonal fluctuations compel the farmers to explore other alternatives. Backyard poultry cum fish pond enterprises provide a reliable source of income for marginal farmers. Silkworm pupae and silkworm skin are considered excellent sources of protein, fats, essential vitamins and minerals (Longvah *et al.*, 2011). A mixture of silkworm larvae and pupae blended in appropriate proportions with other food materials can serve as excellent feed for poultry and fish. Silkworm faecal matter can serve as an effective feed and nutrient source for fish. The feed conversion ratio of silkworm pupae to fish is 2:1 meaning that 2 kilograms of dead pupae can produce 1 kilogram of fish body weight as 80% of cocoon weight is occupied by dead pupae on a weight basis. All by-products from silkworm rearing can aid in better fish yield. Therefore, sericulture in combination with poultry cum fish can be viewed as one of the promising entrepreneurships in augmenting handsome income for the farmers as this combination opens avenues for small scale and commercial agriculture business and hence this type of diversification yields fruitful results (Prein, 2002). On the other hand, waste generated in poultry can be recycled to produce organic manure that can be applied as biofertiliser to boost the growth of mulberry plants which in long run helps the worms to produce good cocoons thus boosting the farmer's resources and income (Kumar *et al.*, 2012; Nigussie, 2018).

Vermicomposting: Vermicomposting is the process of conversion or break down of crop residues of agricultural origin into organic manure with the addition of biological agents known as vermicast or earthworms. Various end products are generated at each stage of sericulture which can be utilised for vermicomposting. The resulting vermicompost can be used for cultivation of various agricultural crops and also for small scale sales. Some additional integrated programmes involving sericulture and other allied agri-enterprises lay emphasis on the following aspects of IFS and sericulture (Baishya *et al.*, 2004).

Plantation and Fruit Crops: Horticulture is a booming industry in India and is particularly significant compared to other agricultural sectors because the fruits grown there are a staple food for everyone and are high in nutritional content. (Kumar *et al.*, 2012). Horticulture offers excellent avenues for generating employment and contributes to the national economy by earning foreign exchange. The production can be increased by increasing the area under cultivation and adopting the idea of mixed cropping. Mulberry (*Morus alba*) the host plant for rearing silkworm *Bombyx mori* has the scope to be planted as an intercrop in between the rows of coconut trees under a mixed cropping system. Mango, coconut, and sapota are major horticulture crops that can be easily combined with the cultivation of mulberry crop (Kerutagi *et al.*, 2019). Thus, farmers do not need to spare their entire cultivated land for mulberry cultivation and can generate additional income by practising a mulberry and horticulture intercropping system. Moreover, the concept of by-product utilisation works well in this direction as the waste generated during silkworm rearing including silkworm litter, undersized and dead worms, left over leaves and twigs can be recycled to produce organic manure when mixed with residual wastes of other agricultural crops. This can serve as an excellent biofertiliser to boost the growth and nutritive value of vegetables and fruits thus minimising the dependency on chemical fertilisers.

Floriculture: Floriculture is an age-old traditional farming activity. It is the branch of horticulture science which deals with the cultivation of flower crops and ornamental plants. In current situations both in India and worldwide, it is one of the promising and economical agriculture businesses. It has received official recognition from the Indian government as a "Sunrise Industry" and has been given a cent percent export-oriented status. Besides floriculture has become one of the major profitable farming activities in the country due to increased demand for a wide variety of flowers. Protected cultivation as well as extensive cultivation and marketing of flowers is becoming a prevalent entrepreneurial venture. Flower cultivation as a supplementary activity helps to improve economic opportunities for rural farmers. This also includes selling in local and distant markets and exporting of cut flowers, live plants and their commercial products such as scents, oils, medicines etc. Farmers can adopt floriculture by starting small nurseries with limited space and time. Sericulture wastes can be recycled into biofertilisers, which can be used to promote the growth of flower crops. Moreover, cultivation of flowers as an intercrop in mulberry fields can also be adopted for additional benefits (Baishya *et al.*, 2004). Cultivation of flowers like marigold and gladiolus or cut flowers can be easily incorporated in IFS (Kumar *et al.*, 2012).

Various Entrepreneurial Opportunities in the Sericulture

It is obvious that the sericulture sector offers a wide range of entrepreneurship opportunities as well as excellent career prospects. The economic fate of rural residents is significantly influenced by sericulture, an agro-based business that integrates well into India's rural system where agriculture is still the dominant industry. Sericulture offers employment options to skilled young people in semi-urban and urban locations in addition to rural communities. By offering chances for income-generating enterprises, sericulture growth would surely result in a prosperous rural area, eliminating poverty and stopping the migration of rural residents to metropolitan areas. An agro-based industry with a very short gestation period, low investment, year-round income flow and high profits. An important feature of sericulture is that the host plant mulberry crop could be grown under a wide range of climatic conditions by small and marginal farmers.

1. **Raising of High Yielding Mulberry Saplings (Mulberry Nursery):** Nowadays, farmers are attracted to sericulture because they can obtain good profits by raising healthy and disease-free saplings of mulberry and marketing them (Figure 2a). Nursery as a business generates high employment opportunities as well as additional income for rural entrepreneurs. Mulberry leaves are the only host plant or feed material for silkworm larvae rearing, hence availability of mulberry leaves is the principal factor to start silkworm rearing. Mulberry plantation garden establishment through raising nursery seedlings is much advantageous than direct planting of stem cuttings in the field, majorly it has better garden establishment rate due to well-developed root system. Nursery seedlings get established rapidly and grow robustly without any stress. Farmers are practicing mulberry garden establishment by planting nursery seedlings realising the better initial establishment of saplings. Hence, raising mulberry seedlings is considered a profitable entrepreneurial activity. Mulberry nursery can prevent the migration of family members to urban areas in search of employment by providing employment opportunities in their own nursery. Around 1.20 lakh seedlings can be raised in one acre of nursery with proper care and management earning ₹240000/- by selling at ₹4/- per seedling. By standard methods, two batches of nursery seedlings can be grown in one year, as the duration of nursery seedlings is 4 months with gap of 2 months earning an income of around ₹480000/- per year.

2. **Production and Supply of Disease-Free Eggs (DFEs):** The timely supply of disease-free eggs is also one of the key factors in the sericulture industry along with feed availability (Figure 2b). A place where commercial mass production of disease-free eggs takes place is called a grainage, from where eggs are supplied to chawki rearing centres and farmers. Careful grainage operations determine the success of silkworm rearing. Therefore, the management of seed production plays an important role in overall returns. Adoption of scientific egg production methods, from the seed crop rearing to egg incubation is necessary to produce quality and disease-free seeds. There is scope for the establishment of private licenced seed production units along with the government to meet the current demand for DFEs. Entrepreneurship in this area is highly profitable and yields a high cost-benefit ratio.

3. **Chawki Rearing Centres (CRC):** Chawki worms are the initial two instars of the silkworms or infant or young age silkworms (Figure 2c). The scientific method of silkworm rearing is a key factor for a

successful harvest of cocoons. Chawki worms are very sensitive to small changes in weather parameters, so it requires well-sophisticated management with technical skill. However, this is very difficult for rural farmers to establish CRCs. Therefore, the CRC business has emerged as a profitable enterprise in the present situation. CRCs have become an essential activity in sericulture and play a key role in the sustainability of bivoltine silk production. Individual business owners, SHGs, and NGOs run commercial CRCs with the sole purpose of ensuring farmers have access to high-quality, healthy worms that will increase cocoon production. Nowadays, farmers rely more on CRCs for chawki worms rather than their own breeding. Chawki rearing of silkworms has emerged as a chief entrepreneurship opportunity in the sericulture industry in sericulture practicing parts of India. For example, an average CRC can hatch 15000 DFLs per month in three different batches of 5000 DFLs per batch in synchronised manner. Marketing 100 DFLs of chawki worms earns a net profit of about ₹500 and it earns approximately ₹65000 to 75000 per month by selling 15000 DFLs (Sujathamma *et al.*, 2018). The establishment of CRCs near cocoon production areas is gaining popularity and is presently working as backward linkage to the farmers. Already many licenced private entrepreneurs are running CRCs on commercial basis in many parts of southern states like Karnataka, Andhra Pradesh and Tamil Nadu.

Economics of Chawki Rearing Centre

CSRTI, Mysore has designed plan for ideal CRC building with capacity of rearing 5000 DFLs per crop and totally 32 crops in a year (1.6 lakh DFLs per annum).

Total Non-Recurring Costs	₹1459250 (Depreciation for 1 year is ₹75683)
Recurring Costs	₹1762664
CRC Owners Profit	10% of total cost

1. **Cocoon Production:** Silkworms are reared for cocoon production with coordination of series of events such as establishing mulberry garden, disinfecting and cleaning the building and equipment purchasing DFLs, silkworm rearing, maintaining hygiene, caring for the worms during moulting, mounting worms for spinning and cocoon harvesting (Figure 2d). Under ideal conditions, the cycle takes 25 to 30 days from the day of hatching to cocoon harvesting and the cycle can be completed within 21 to 22 days by using chawki worms instead of eggs. The 5th instar larvae stop feeding and spin the cocoon with silk protein called fibroin. This

occupation has emerged as a profitable venture within a short period of time simultaneously with regular cultivation of crops, especially in favourable regions. The scientific method of rearing produces about 8 to 9 crops per year (by dividing one acre into two plots), it has become the main source of income for many farmers. On average 100 DFLs produce about 70 to 80 kg of cocoon yield. The present average price of cocoons is between ₹450/- and ₹500/- per kg (Bivoltine). A one-acre mulberry garden facilitates rearing around 1800 DFLs per year with an expected cocoon yield of about 1440 kg. Approximately ₹576000/- and ₹400000/- in gross and net income respectively can be generated per year (Sujathamma *et al.*, 2018).

2. **Host Plant Cultivation (Moriculture):** The production of good quality feed material is the key step in silkworm rearing (Figure 2e). Apart from being used as feed material, plants have many uses in economical point of view. Only the leaves are used in the case of silkworm rearing and leftover foliage, fruits, wood material etc., are considered as by-products. So, there is a lot of entrepreneurial scope for farming families in mulberry plant cultivation.

Entrepreneurial Prospects Using By-products Generated in Different Stages of Sericulture

a) **Mulberry Leaves:** Mulberry tea made from leaves of mulberry has health benefits for the proper functioning of kidney and liver and also improves hearing problems. The leaves can also be used as quality animal feed as they are rich in Ca, P, Mg, and vitamins such as B, C, K and other antioxidants. The medicinal benefits of mulberry leaves include diuretic properties, blood sugar and blood pressure reducing effects. The leaves are also used as anticancer agent and weight loss agent. So, the leaves are used in the preparation of various types of medicines and herbal supplements (Nazim *et al.*, 2017).

b) **Mulberry Fruits:** Mulberry trees produce berries with delicious flavours, outstanding nutritional value, and several health advantages. Mulberry fruits are becoming more and more popular over the world. The fruits are rich in vitamins and minerals, especially vitamin C and iron. They are used in the preparation of jam, jelly, pickle, squash, juice, wine, vinegar etc. and different cosmetic products. Also, they are used as a natural colouring agent and its roots are an excellent raw material for the pharmaceutical industry as they contain high flavonoid and phenol content.

c) **Use of Cocoons for Cosmetic Purposes:** Silkworm cocoons can be used to gently and precisely exfoliate the skin, resulting in healthier, softer skin. The natural collagen, proteins, and amino acids found in silk cocoons help keep the skin smooth and moisturised, eradicate blemishes, and act as anti-aging agents. Lily Silk Beauty Cocoon is a specialised silk store, a new venture that has started manufacturing and marketing beauty products of good quality. It was first started in China.

d) **Craft Items from Cocoons:** Designing handcrafted products such as flowers, bouquets, garlands, greeting cards, and other artistically styled objects, is known as cocoon crafting. Defective cocoons, double cocoons, stained cocoons (dead cocoons), deformed cocoons, thin end cocoons, pierced cocoons, multi-end cocoons, flimsy cocoons etc, can be used to produce attractive handcrafted items or products which will avoid the wastage of cocoons.

e) **Reeling and Spinning**: Reeling and spinning are two activities for the production of raw silk from cocoons. Reeling is done in the case of continuous silk filament. The price of 1 kg of mulberry, *eri* and *muga* raw silk is about ₹3000-3500/-, ₹2250-2800/-, and ₹16500-25000/-, respectively. There is always a demand for quality silk. Providing training to rural youth regarding improved techniques in rearing aids in quality silk production and also reduces unemployment.

f) **Twisting:** The raw silk extracted from the cocoon is to be twisted before weaving. The process of twisting imparts spiral turns into the raw silk. The twist holds together the fibres and also increases the strength of the yarn. The twist of the yarn plays a key role in determining the characteristics of the finished cloth. Though the initial investment is high to establish a twisting unit, a substantial profit can be earned from twisting raw silk.

g) **Dyeing and Printing**: The method of imparting colour to fibre, yarn or fabric is termed as dyeing. Printing on textiles enables the creation of colourful fabrics. Human beings are always attracted to colourful objects. Dyeing and printing add value to silk material. Nowadays, silk dyeing with natural dyes is gaining popularity not only in the foreign market but also in the domestic market. These two industrial activities have the potential to become profitable enterprise in silk-growing states in the country.

h) **Weaving**: Silk weaving is a family occupation, a traditional activity passed down from generation to generation through the educational learning process within the family as a source of income generation. Silk weaving in handloom with ethnic designs has a high demand in both domestic and foreign markets.

i) **Finished Product/Fabric:** Silk fabric is known as the queen of all fabrics. A major portion of the silk produced in India is consumed by Indian people in different forms. Mekhela-chadar, saree and salwar-suits are some of the finished products of mulberry silk popular in different parts of the country. Mekhela-chadar, saree, gamocha, tie, purse etc. can be prepared from *muga* silk whereas shawl, mekhela-chadar, socks can be prepared from *eri* silk. Muga and eri silk yarn and fabric have a high demand in the international market. This is also a profitable enterprise in the age of fashion.

Suggestions for Boosting Entrepreneurship in Sericulture

- Facilities for infrastructure development at the farmers level,
- Popularizing dual cropping and digital marketing,
- Incentives for farmers,
- Infrastructure development/upgradation support from the government,
- Introducing a crop insurance scheme for cocoon production,
- Marketing facilities for Indian silk and sericulture products at international level,
- Public-Private partnership in the industry,
- Improving research and development in research institutes through capacity building and skill development,
- Providing financial facilities from the government,
- Educating the youth through training, workshops, etc.,
- Providing financial support for startups in sericulture from MSME schemes, and
- Large-scale establishment of the above-mentioned opportunity fields in sericulture.

Sericulture based small scale dairy farming

Sericulture based fish cum poultry farming

Mulberry based intercropping system

Vermicomposting using silkworm bed residues

Fig. 2: Entrepreneurial Opportunities in Sericulture-based Integrated Farming System

3(a). Mulberry Nursery

3(b). Silk Worm Egg Production (Grainage)

3(c). Chawki Rearing Centre

3(d). Silk Worm Rearing

3(e). Mulberry Production Field

Fig. 3: Entrepreneurial Opportunities in Sericulture

Mulberry fruits | Cocoon crafting | Cocoon crafts

Hand weaving | Dyeing of silk | Ready made silk sarees

Fig. 4: Entrepreneurial Opportunities with Sericulture By-products

Success Story of Young Boy who left Mumbai and Succeeded in Sericulture-based IFS

A young boy named Praveen, a resident of Vaggarahalli village, Channarayapattana taluka, Hassan district of Karnataka has a family land holding of 5 acres. He left his schooling in 10th and started working in Bengaluru and Mumbai due to family debts for one year. Due to low salary and being away from home he left Mumbai and started his new venture by purchasing a cow worth ₹9500/- 20 years ago. He led his family by selling milk and later he increased the number of cows to 5 and started a small scale animal husbandry with selling of 50-60 litres per day. After years, he started sericulture by constructing a silk worm rearing house and planted mulberry in half an acre with a capacity of 60-100 DFLs per crop. He has since increased the area under mulberry to 2 acres and also increased the number of DFLs per month, now earning around 6-7 lakhs per annum from sericulture as a major income-earning activity. He started IFS with the advice of a nearby agriculture expert in KVK, incorporating backyard poultry, goat farming, animal husbandry and planting coconut, arecanut in an intercropping system, coffee, pepper, cardamom, fruit trees like mango, banana, guava, sapota ramphal, lakshanphal and fodder trees all along the borders as well as in between for both family needs and marketing. He was earning a profit of ₹25000 to ₹30000/- per annum from poultry, a weekly income of ₹1500/- from selling desi eggs at the rate of

₹15 per egg, ₹2.5 lakhs per annum from animal husbandry, ₹2-3 lakhs from coconut and arecanut, ₹30000/- profit per annum from small scale goat rearing and a major profit of ₹6-7 lakhs per annum. Besides he was growing ragi, pulses and vegetables fully organically using FYM from animal husbandry and green fodder as feed for goats. Finally, he was earning a profit of around ₹13 lakhs per annum from all the activities. He has built a house worth ₹35 lakhs, purchased a tractor and is leading his family without any deficiencies. He attributes his success to the sericulture-based integrated farming system which has provided employment to his family members and many others.

Conclusion

From this chapter, it may be inferred that there is ample scope to develop entrepreneurship in the sericulture sector. The present youth have entrepreneurial ideas, but only a very small percentage of youth have the opportunity to transform such ideas into successful enterprises. The ability of youth to transform their inventions into businesses is critical to the future of start-ups, especially for those who belong to small and marginal farming communities. The sericulture-based integrated farming system has provided many innovative and creative entrepreneurship opportunities as sericulture is a multifaceted industry. Presently, rural youth should focus on startups based on sericulture-based IFS by utilizing financial assistance from both central and state governments in sericulture fields.

References

Baishya, A., Pathak, A.K., Bhowmick, B.C. and Ahmed, S. (2004). Pre-dominant farming system and alternative in Assam in alternative farming systems. Proceeding of a Symposium held at PDCSR, Modipuram, Meerut on 16-18 September, pp. 228-237.

Hamid, N. and Bhat, A. (2021). Seri-Business: New Emerging Entrepreneurship Model in Sericulture. AgroScience Today, 2(5): 0145-0148.

Kerutagi, M.G., Talavar, M. and Pavitra A.S. (2019). Impact of horticulture-based integrated farming system on farmers' income and welfare in Northern Karnataka. Journal of Pharmacognosy and Phytochemistry. 8(3): 1010-1019.

Kumar, S., Dey, A., Kumar, U., Chandra, N. and Bhatt, B.P. (2012). Integrated farming system for improving agricultural productivity. Status of Agricultural Development in Eastern India, Publisher: ICAR Research Complex for Eastern Region, pp. 205-230.

Longvah, T., Mangthya, K. and Ramulu, P.J.F.C. (2011). Nutrient composition and protein quality evaluation of eri silkworm (Samia ricinii) pre-pupae and pupae. Food Chemistry, 128(2): 400-403.

Nagaraju, Y. and Raghavendra, N. (2016). Improve Livelihood Security and Employment Generation through Integrated Farming System of Scheduled Caste Farm Families in CB Pura District of Karnataka. International Journal of Science and Research. 5(8): 1419-1421.

Nazim, N., Buhroo, Z.I., Mushtaq, N., Javid, K., Rasool, S., and Mir, G.M. (2017). Medicinal values of products and by products of sericulture. Journal of Pharmacognosy and Phytochemistry, 6(5): 1388-1392.

Nigussie, A. (2018). Livestock and Fisheries Research Strategies. Livestock and Fisheries Research Strategies Poultry, Fisheries, Apiculture and Sericulture (2016-2030). Ethiopian Institute of Agricultural Research, pp. 01-227.

Prein, M. (2002). Integration of aquaculture into crop–animal systems in Asia. Agricultural Systems. 71: 127-146.

Saikia, M. (2021). Entrepreneurship development in sericulture sector of Assam. Just Agriculture, 1(12): 10-13.

Savthri, G., Sujathamma, P. and Kumari, N.V. (2009). Sericulture for Sustainable Rural Development in Indian Perspective. Rural Development, pp. 141-144.

Sujathamma, P., Savithri, G. and Neeraja, P. (2018). Sericulture – A Potential Agro-based Enterprise. International Journal of Research in Science and Engineering, 7(7): 203-207.

11

Role of Beekeeping for Income Generation Through Integrated Farming System

***Kumari Asmita*[1], *Sudhanand Prasad Lal*[2] and *Richa Kumari*[3]**

[1]*AICRP on Women in Agriculture, College of Community Science, Dr Rajendra Prasad Central Agricultural University, Pusa, Samastipur, Bihar*
[2]*Department of Extension Education (PGCA), Dr Rajendra Prasad Central Agricultural University, Pusa, Samastipur, Bihar*
[3]*Home Science, Lalit Narayan Mithila University, Darbhanga, Bihar*

Abstract

Beekeeping is the greatest rural people supported enterprise for powering the financial condition of the entire community that can be applied by everyone. Beekeeping does not require any grave early funding or nonstop work, so it can be integrated with every additional farming practice. It offers a variety of products such as honey, pollen, and wax etc. Honey bees can be used as an additional very important agro effort for enhancing crop efficiency. Today, in our country farmers and entrepreneurs are attracted to beekeeping practices because of its increasing requirement and desire for earning high turnover by obtaining huge making and gathering honey with the help of the latest, precise and technical methods of honey collection. Apiculture and honey-producing practices are one of the socio-economic activities, which are generally friendly to forests and the environment too. For an extensive period, agriculture has recognized the worth of pollination by bees. Yet garden beekeepers witness impressive progress in their yields of gardens in fruits, flowers and vegetables. About 16 lakh people are directly or indirectly involved in bee-keeping and its related enterprises in India. The most important honey producing states in India are Punjab, Haryana, Bihar, Himachal Pradesh, Uttar Pradesh and West Bengal. The role model of beekeeping, 16 Integrated Beekeeping Development Centres (IBDCs) has been sanctioned individually apiece in the States of Haryana, Delhi, Bihar, Punjab, Madhya Pradesh(MP), Uttar Pradesh(UP), Manipur, Uttarakhand, Jammu & Kashmir(J &K), Tamil Nadu, Karnataka, Himachal Pradesh(HP), West Bengal(WB),

Tripura, Andhra Pradesh(AP) and Arunachal Pradesh. There is enough scope for taking up beekeeping on a commercial scale. But the low price of honey and its by-products is rated as the most serious constraint by the beekeepers.

Keywords: Agriculture, Apiculture, Beekeeping, Income generation, Integrated Farming System.

Introduction

Beekeeping is a vital component of agriculture and rural development programmes in various Asian countries. The great role of beekeeping in supplying nutritional, economic and ecological safety to rural areas of Asia cannot be ignored, as it has always been associated with their cultural and natural traditions. Beekeeping is a landless activity and does not compete with any other resource-intensive elements of farming systems. Beekeeping is an additional income-generating activity at the household level. In the crop fields, pollination activities of honeybees are essential, as they help boost production. Apart from crop production and horticulture, dairy production and other businesses in India promote agricultural production, which are helpful in promoting the economy of India, one of them is apiculture. Apiculture is a very less expensive business and can be used by landless farmers to all groups of farmers for increasing their income. By selling various types of substances like honey, beeswax, etc. in the market, they get profits in millions. In the beekeeping process, bees carry out the pollination process in flowers, which directs to an extra 1/4th increase in crop production that helps in raising agricultural and horticultural production. Beekeeping is an ideal rural business that any individual can start because it increases the overall economic stability of the community. It generates a wide range of commodities, such as honey, pollen, and wax, in addition to pollination. Honey bees are the primary pollinators that can be easily produced in large numbers if and when needed. Honey bees can be used as a different essential agro input, similar to fertilizer, seeds, pesticides, irrigation, etc., to increase crop yield. Farmers' income and, ultimately, their standard of living are improved by the sale of beekeeping goods and the increase in crop yield brought on by bee pollination. There is enough scope for taking up beekeeping on a commercial scale. Beekeeping has a special benefit in India where the majority of farmers are small or marginal land-holders and a large area is under horticultural crops.

Meaning of Apiculture (Beekeeping)

The word 'apiculture' derives from the Latin word '*apis*' which means bee. The bee genus *Apis* is referred to as "apiculture." The two most common types of honey bees kept are *Apis mellifera* and *Apis cerana*. Beekeeping, also known as apiculture, is the process of protection and management of honey bee colonies, generally utilising man-made hives. ***"Apiculture is the technical process of rearing honeybees."*** The process of gathering honey from the forests has been in survival for a long time is known as apiculture. Honey bees change nectar of flowers into honey and stock them up in the combs of the hive, increasing market possibility for honey and its different products have also increased bee keeping growing like an applicable enterprise.

Advantages of Beekeeping

- It requires the least amount of time, money and resources.
- Beekeeping has encouraging biological effects. Bees play a significant role in the propagation of several flowering plants, boosting the productivity of certain crops like sunflowers and a variety of fruits.
- Maximum demand of honey and wax in the marketplace.
- Honey is a mouth-watering and greatly healthy food. By raising bees in boxes and producing honey at home, chances of loss of wild colonies of bees can be prevented.
- Beekeeping enterprise can be started by a single person or a group of two or more people.

Integrated Farming System

Integrated farming is a complete farm management system which means to deliver more sustainable agriculture. Farming involves harnessing solar energy to produce cost-effective plant and animal products. The system involves a set of practices and procedures systematised into a useful unit. The term "Integrated Farming System" (IFS) refers to a set of inter-reliant, related, and commonly interconnected production approaches which are based on a tiny numeral of crops, animals, and related contributory industries and proposed to exploit the competence of nutrients in every system and reduce the harmful environmental consequences of these entrepreneurs.

Goals of Integrated Farming System

- Exploitation of yield of every constituent enterprises
- Poverty reduction and secure income
- Renovation of systems production
- Accomplish agro-ecological stability

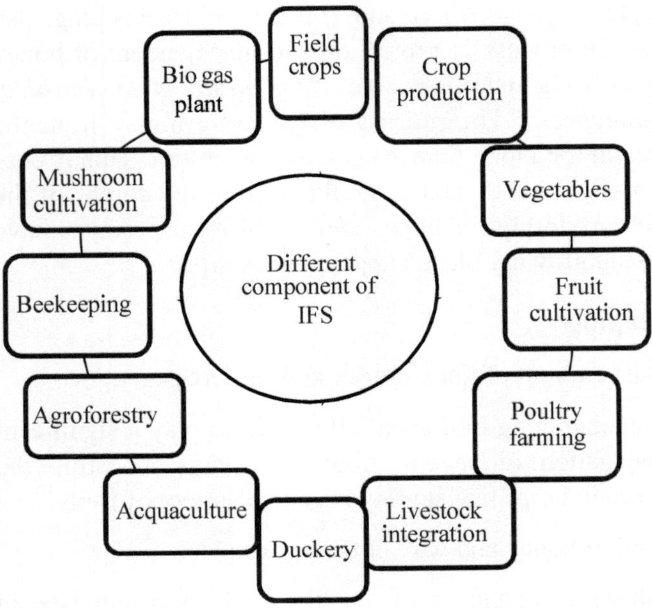

Fig. 1: Different component of IFS

Significance of Beekeeping

Beekeeping has numerous benefits for the development of farmers, the nation and the environment in general. There are different benefits of beekeeping:

- It boosts the value and extent of produce e.g. fruits, vegetables, oilseeds, pulses, etc.
- It facilitates in sustaining biodiversity.
- It gives an essential contribution to agriculture.
- It is a low cost technology.
- It is a good source of income and creates employment opportunities.
- It helps in the sustainable development of the environment as well as agriculture.
- It helps in increasing the earnings of farmers and improving their economic status.

Table 1: Different Types of Plants for Apiculture Farming in India

Types of Plants used for beekeeping	Forest/Region
• Citrus • Coffee • Coriander • Jamun • Litchi • Mohul • Prosopis • Rubber • Sandal wood • Soapnut	• Himachal Pradesh • Karnataka • West Bengal • Mahabaleswar • Bihar • Madhya Pradesh • Kutch • Kerala • Kadappa forest • Coastal Karnataka & Andhra Pradesh

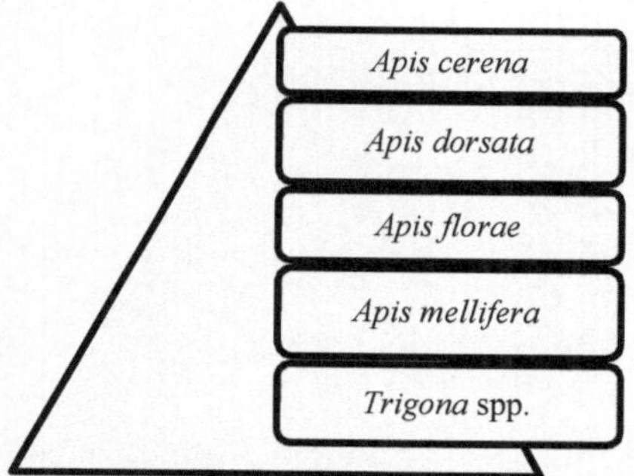

Fig. 2: Honeybee species for beekeeping in India

Present Scenario of Honey Production in India

The significance of beekeeping as an element of the Integrated Farming System in the nation, the administration endorsed allocation of Rs.500 Crore for the National Beekeeping & Honey Mission (NBHM) for three years (2020to 2023), which is declared as a part of the Atma-Nirbhar Bharat scheme. The main aim of NBHM is to promote and develop scientific methods of beekeeping in the country to achieve the goal of the **'Sweet Revolution'** which is being executed through the National Bee Board (NBB).

Objective of NBHM

1. To encourage the integrated growth of the beekeeping industry to generate income and employment opportunities for farm and non-farm households.
2. To improve agricultural/ horticultural production by increasing infrastructural capacities, containing the conception of Integrated Beekeeping Development Centres (IBDCs)/CoEs, Api-therapy centres, bee disease diagnostic labs, honey testing labs, bee breeders, etc.
3. To empower women through beekeeping.

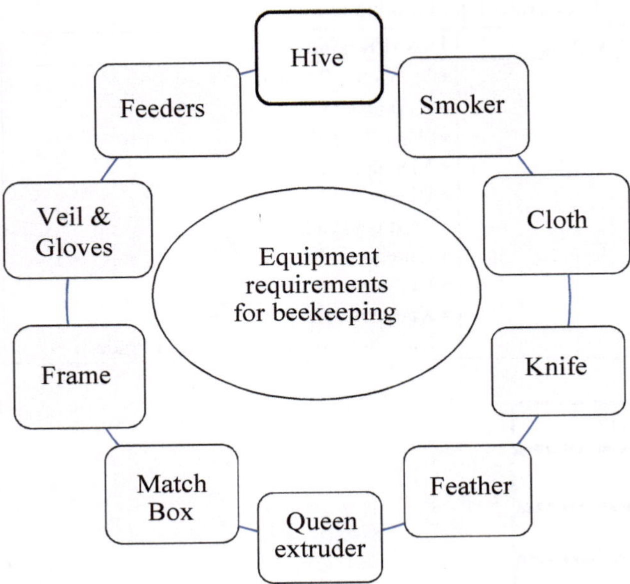

Fig. 3: Equipment requirements for beekeeping

Beekeeping in Bihar

Table 2: Brief details of Mr. Ajay Sahni

Sl.No.	Details	Specifications
1.	Agro climatic zones	Middle Gangetic Plains
2.	State	Bihar
3.	Region	Agro climatic zone-I (Northern West)
4.	District	Samastipur
5.	Block	Kalyanpur
6.	Village Panchayat	Hajpurwa
7.	Village	Bakhtiyarpur
8.	Latitude and longitude	25.90840624024427, 85.83030999865744
9.	Farmer name	Ajay Sahni
10.	Date of Birth	10/04/1982
11.	Present Age	40
12.	Education	Up to 9th class
13.	Occupation	Apiculture/beekeeping
14.	Inception of beekeeping	Since 2012
15.	No. of Wooden Boxes	150+ Boxes
16.	Area of Beekeeping	Bihar, Jharkhand, UP, MP
17.	Maximum output	6-7 Lakh/Year
Source of Village: (ONEFIVENINE, 2021)		

Apiculture (Beekeeping) in Bihar is enhanced by an extremely expanded, rich bee-flora and good ecological conditions. The litchi honey produced in Bihar has a superior taste, colour and essence compared to honey produced in other states due to differences in ecological conditions. Litchi honey is very popular and highly demanded at the national and international levels. Honey bees act as outstanding pollinators and thus expand the yield of a variety of agricultural crops. In Bihar, migratory beekeeping has been expanded by the All India Coordinated Project (AICRP) on Honeybee, Pusa centre since 1993 which haves proven valuable for producing diverse forms of honey in different areas, resulting in a significant enhancement of yield and colony life. By adopting migratory beekeeping, beekeepers can produce different honey flows and increase their income.

Fig. 4: Beekeeping in Bihar by Ajay Sahni

This is the story of Mr. Ajay Sahni who is lives in Bakhtiyarpur village of Hajpurwa panchayat, Kalyanpur block in Samastipur district of Bihar. He was born on 10th April 1982. He has passed 9th class. Mr. Ajay started beekeeping 10 years ago (in 2012) after getting training from Dr Rajendra Prasad Central Agricultural University, Pusa, Samastipur, Bihar and his areas of beekeeping is Bihar and other states also like Utter Pradesh (UP), Madhya Pradesh (MP) & Jharkhand. He has been keeping over 150 beehives and makes a net profit of about Rs 6-7 Lakh per year. This is the example we have collected from the field. He is a skilled beekeeper who also trains other people in his village and nearby villages in beekeeping and helps them increase their profits. He has become an inspiration in the beekeeping field in his community.

Diagram of Honey Processing Process

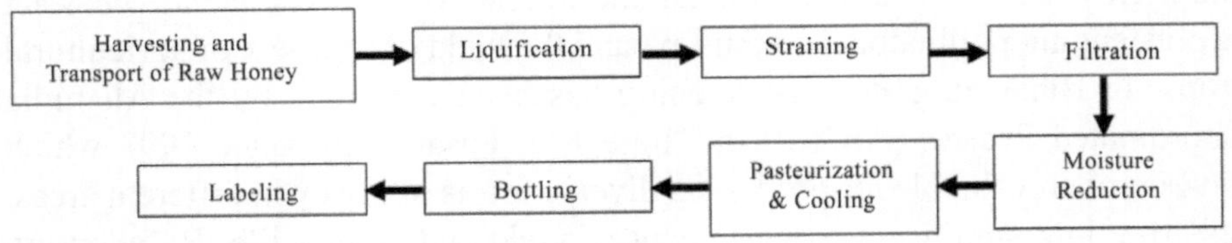

Different products of beekeeping and their uses

1. **Honey:** It has been used as an ordinary sweetening agent. It has immense use in the pharmaceutical industry and is also considered a medicine by *Ayurveda*. It is popularly used as a household cure for cough and a preferred consumable for people on diet, in the preparation of confectionaries and is also used for consumption with bread, bakeries and cakes etc. It is also used to prepare antiseptic and skin care lotions and creams. Good quality honey has a high demand in the international market and has the potential to generate substantial foreign exchange for the country. Honey is marketed in several different forms: **liquid honey, comb honey, and creamed honey**.

2. **Bees Wax:** It is one of the expensive products of beekeeping production that is secreted by worker bees. Wax is used in making quality candles, cosmetics, agriculture, art and industry.

3. **Bees Venom:** It is clear liquid with a sharp bitter flavour and pungent odour. It is used in the treatment of asthma, neurosis and migraines. It enhances haemoglobin levels and reduces cholesterol and blood pressure.

4. **Propolis:** It is a sultry, tacky resinous substance collected by bees. It can be used as a curing agent and is useful against itching. It has diverse qualities that could help improve human health. It can be used by people on the skin, as a mouthwash etc.

5. **Royal Jelly:** It is a creamy white color food that nourishes queen bee larvae and is incredibly rich in proteins and fatty acids. It is not considered important in human nutrition but it has been used in cosmetics as skin conditioners and facial creams.

6. **Bee Pollen:** It offers nutritional food for bees. It is a unique combination of bee secretions, enzymes, honey, wax, and flower pollen. Bee pollen may have various benefits for human beings, however gathering it may adversely affect the well-being of the bees and even destroy the colony. People can use it in various food products, such as a topping on cereals, salads, yogurt, in smoothies and other related drinks.

Challenges

Development of beekeeping and honey production units in India are facing a lot of challenges because of diverse internal and external causes, such as:

1. Short age of fruits and vegetables, floral scarcity/dearth periods.
2. Lack of crop diversification.
3. Excessive use of chemicals in agriculture and other crops.
4. Climate change.
5. Lack of technical management practices.

Conclusion

Beekeeping is an agricultural activity carried out by farmers and labourers including those without land in rural areas as part of an Integrated Farming System (IFS). It is helpful in pollination of crops, increasing profits for beekeepers through higher crop yields and providing valuable products such as honey, bee wax, bee pollen, etc. Overall, apiculture is the most widely practiced activity by farmers in the country and is an essential component of smallholder farming systems. It is also a leading income-generating activity in rural areas. Beekeeping doesn't require much money, large amount of land and very deep complicated technical knowledge. Therefore, it is a cost-effective practice for small farmers with limited resources, including women and youth. Therefore, if this off-farm practice is promoted with special consideration for women and new generation youth, it will be a significant step towards fulfilling the country's plan of achieving the goal of the Agricultural Growth Program Phase Two (AGPII) which emphasises the maximum involvement of women and youngsters. Apiculture is an economically profitable enterprise and can be a useful enterprise as it can add income and improve the economic condition of farmers.

Then it decreases the perception that apiculture is solely a men's work with only a small number of women having concern to start the enterprise. Bees are very essential for maintaining a good environment because without bees, pollination is impossible. Pollination is necessary for the production of new plants, animals and ultimately life on earth.

References

Baba, S. (2022). Bee production process, Academia ©2022. https://www.academia.edu/5505451/Bee_production_process

Bee Keep club (2022).What is beekeeping. The beginner's introduction. https://beekeepclub.com/getting-started-beekeeping/what-is-beekeeping-the-beginners-introduction/

BYJU'S (2022). Apiculture: Beekeeping, Biology Article, Copyright @think & learn Pvt. Ltd. https://byjus.com/biology/apiculture-beekeeping/

BYJU'S EXAM PREP (2022). National Beekeeping & Honey Mission (NBHM). https://byjus.com/free-ias-prep/national-beekeeping-and-honey-mission-nbhm/

Chamola, D. &Behera, B. (2019).Business Model Promoting Bee Keeping, Deutsche GesellschaftfürInternationaleZusammenarbeit (GIZ) GmbH, NABARD, August 2019, New Delhi. https://birdlucknow.nabard.org/wp-content/uploads/2020/03/11-Promoting-Bee-Keeping-DNPPCL.pdf

Dafar, A. (2018). Review of Economic and Ecological Importance of Bee and Bee Products in Ethiopia. Journal of Animal Husbandry and Dairy Science, 2(2): 18-26. https://www.sryahwapublications.com/journal-of-animal-husbandry-and-dairy-science/pdf/v2-i2/3.pdf

Esakkimuthu, M. and Kameswari, VLV. (2017). Entrepreneurial Potential of Small-scale Beekeeping in Rural India: A Case in Kanyakumari district, Tamil Nadu, Tropical Agricultural Research, Vol. 28(4): 411-424. https://www.researchgate.net/publication/329655898_Entrepreneurial_potential_of_small_scale_beekeeping_in_rural_India_a_case_in_Kanniyakumari_district_Tamil_Nadu

Godifey, G. &Tassew, A. (2016). Importance of Integrating Beekeeping with Closure Areas in Ethiopia: Status and Future Prospects, Journal of Biology, Agriculture and Healthcare, 6(3):30-41. https://core.ac.uk/download/pdf/234661884.pdf

Khadi and Village Industries Commission (2022). Operational Guidelines for the Pilot Projects of Beekeeping Activity under Agro-Based & Food Processing Industries (ABFPI) vertical of GramodyogVikasYojana(GVY), Ministry of MSME. https://msme.gov.in/sites/default/files/Beekeeping.pdf

Kishan, M., Aruna, R., Mishra, G. &Srinivasan, M.R. (2017). Beekeeping in India, ResearchGate, Industrial Entomology published on February, pp 35-66.DOI: https://www.researchgate.net/publication/313692521_Beekeeping_in_India

Lal, S.P., and Jha, S.K. (2018). Benefit-cost and constraints analysis of beekeeping in Green revolution province of India, published on January-February, Bhartiya Krishi Anushandhan Patrika, 33(1):38-42.

Malik, A. and Azad, A. R. (2020). Economic Importance and Utilization of Apiculture, published on November, in Just Agriculture, Madhya Pradesh. https://justagriculture.in/files/magazine/nov/004%20Economic%20Importance%20and%20Utilization%20of%20Apiculture.pdf

Mouhammad, B. (2018). Income Generating Activities (IGAs), Food and Agriculture Organisation of the United Nations, January 16. https://fscluster.org/sites/default/files/documents/igas_presentation_fssm_16_01_18_0.pdf

NDA (2013).Income generating activities for the poor in South Africa: The role of Civil Society Organisations, Publications Policy Brief, March 01. https://www.nda.org.za/publication/income-generation

ONEFIVENINE. (2021). Bakhtiyarpur Map. URL: http://www.onefivenine.com/india/villages/Samastipur/Kalyanpur/Bakhtiyarpur

Pande, R., Ramkrushna, G. I., Verma, P. & Shah, V. (2020).Role of Honey Bees for Income Generation in Farming System, Biotica Research Today- An international E-Magazine, 2(11):1122-1125(Nov.2020):https://biospub.com/index.php/biorestoday/article/view/541

Payeurs, S. (1996). "IV - Income generating activities (IGAs)" in: "Mission report: development of activities for women communities in Jordan and Syria". Rome: Food and Agriculture Organisation of the United Nations, Accessed, June 12, 2020. https://socialprotection.org/learn/glossary/income-generating-activities

Perez, A. and Johnson, J. (2021).What are the benefits of propolis, Medical News Today, medically reviewed on March 10. https://www.medicalnewstoday.com/articles/propolis

PIB (2021).National Beekeeping & Honey Mission, posted on February 11, Ministry of Agriculture & Farmers Welfare, Government of India, Delhi. https://pib.gov.in/PressReleseDetailm.aspx?PRID=1697113

Pratap, U. &Verma, L.R. (2022). Asian Bees and beekeeping: Issues and Initiatives. https://lib.icimod.org/record/22855

Richter, A. and Johnson, J. (2021). Bee pollen: What to know, Medical News Today, medically reviewed on May 18. https://www.medicalnewstoday.com/articles/bee-pollen

Singh, V., Verma, D. K. and Chauhan, D. (2017).Beekeeping Technology and Honey Processing: Emerging Entrepreneurship for Rural Areas, First edition, Apple AcademicPress. https://www.taylorfrancis.com/chapters/edit/10.1201/9781315194677-1/beekeeping-technology-honey-processing-emerging-entrepreneurship-rural-areas

TIME IS (2019).Honey Processing, uploaded on September 2019, Department of Science & Technology, GOI, FICCI. https://www.techno-preneur.net/technology/project-profiles/food/Honey.htm

Vedantu (2022).Apiculture Beekeeping, Biology, Study Material, Book Online Demo. https://www.vedantu.com/biology/apiculture-beekeeping

12

Seaweed Farming – A New and Viable Step Towards Sustainable Entrepreneur Development in Integrated Farming System

Kajal Sengupta and Anannya Dhar

Division of Agronomy, Faculty of Integrated Rural Development and Management (IRDM), School of Agriculture and Rural Development Ramakrishna Mission Vivekananda Educational and Research Institute [Deemed to-be-University], Narendrapur Campus, Kolkata, West Bengal

Abstract

Seaweed (macrophytic marine algae) farming is a recent trend in agriculture that has the potential to bring a sustainable revolution in various fields. Seaweeds are nontoxic, biodegradable, non-polluting, and non-hazardous to humans and the environment. Regarded as the 'Medicinal food of the 21st century", seaweeds are used as food for humans and feed for animals. The bio-stimulatory effect of seaweed increases the growth, yield, and quality of agricultural produce. India is blessed with a shoreline of about 7,500 km which can be utilised as seaweed cultivation zones. Seaweed farming is emerging as an alternate occupation for the fisherman community and also promotes women's empowerment. Integrating seaweed farming with other allied enterprises like fish/shrimp farming and Mollusca farming will lead to an increase in net return and also make the system sustainable by using the waste of one unit as the input for others thus reducing the level of pollution. The income obtained from seaweed farming along with integration with other allied enterprises is good and sustainable. Besides, the ecosystem services rendered by seaweed farming like scrubbing excess nutrients from the water and producing oxygen prove to be an additional benefit. Farmers should be made aware of the benefits of seaweed farming and its integration into marine aquaculture systems to promote sustainable entrepreneurial development. Seaweed farming has proven to be a sustainable green business model for the present and upcoming generations.

***Keywords*:** IMTA, Seaweed farming, Sustainable aquaculture, Eutrophication, Nutrient scrubber.

Introduction

Seaweeds are referred to as the wonder plants of the sea, macrophytic marine algae which are a primitive type of plants lacking true roots, stems and leaves. They usually grow in shallow coastal and marine water. Based on the occurrence of pigments, seaweeds are classified into three groups: Green seaweeds (*Chlorophyta*) having the pigments chlorophyll ***a*** and chlorophyll ***b***; Brown seaweeds (*Phaeophyta*) having the pigment fucoxanthin; and Red seaweeds (*Rhodophyta*) having the pigment phycoerythrin. Seaweed is a climate-smart algae that absorbs significant amounts of carbon, reduces acidification of the oceans, and acts as a nutrient scrubber. Therefore, seaweed is regarded as the '*medicinal food of the 21st century*' as they contain unique bioactive compounds, vitamins, minerals, and trace elements and are used in making pharmaceutical capsules for the treatment of cancer, goitre, cardiovascular surgeries, etc.

The input requirements for seaweed farming are very minimal as they do not need fresh water for irrigation, arable land, or nutrients. Former President of India, Dr. A.P.J. Abdul Kalam stated "*seaweed cultivation neither requires land nor irrigation water nor any fertiliser; instead, it yields fertilisers, which will be used in land-based crops*". Besides being good nutrient scrubbers, they take up dissolved nitrogen, phosphorous, and carbon from the water and increase the biodiversity of the sea by giving shelter to a variety of organisms. Seaweeds are known to play a crucial role in reducing the adverse effects of climate change.

The diversified application of seaweeds includes nutritious food for humans, palatable feed for livestock, poultry, and fish, biostimulants for boosting agricultural crop production, in the cosmetic and pharmaceutical industry, and the production of agar, carrageenan, and agarose. Seaweed extracts contain many phytohormones which are known to increase the uptake of plant nutrients from the soil, reducing the external doses of fertilisers and helping plants adapt well to adverse climatic conditions, making the system of production sustainable.

India is blessed with a shoreline of about 7,500 km which can be utilized as seaweed cultivation zones. The fisherman communities dwelling in these coastal areas can be used as manpower for seaweed farming. Seaweed farming will provide an alternate source of occupation for the fisherman communities. It also provides employment to women thus promoting women's empowerment and making the women financially self-sufficient. According to the estimates of ICAR Central Marine Fisheries Research Institute (CMFRI), our country produced around 34,000 tonnes of seaweed in the year 2021, which was far behind in terms of seaweed production (We have the potential of producing 9.7 million tonnes of seaweed per year on a wet weight basis).

Integrating seaweed farming with other allied enterprises like fish farming and mollusc farming will lead to an increase in net return and also make the system sustainable by using the waste of one unit as the input for others thus reducing the level of pollution. In integrated multi-trophic agriculture (IMTA), aquatic species of varying trophic levels are used and wastes or co-products of one cultured species may be used as inputs for the other species. The fed species (shrimp/fish/) releases particulate organic matter (POM) into the water due to unconsumed food and faeces production. The fed species also produces DIM (3, 3-diindolylmethane) as nitrate, phosphate, and carbon dioxide due to metabolism. The molluscs which act as a POM filter, absorb the POM and boost their diet. The seaweeds take the DIM and use it for growth and development. Also, the seaweed produced can be used as feed for livestock and nutrients for agricultural plants.

Seaweed cultivation in India faces various challenges which include a lack of diversification in the use of seaweed products, a lack of awareness among farmers to integrate seaweed into their farming systems, a lack of availability of quality seed products, a lack of proper marketing facilities and other environmental constraints such as climate change. The problems need to be addressed to improve the production system and promote sustainable farming of seaweeds with good revenue generation.

Entrepreneurship Potential for Seaweed Cultivation in India

India is blessed with around 7,500 km of coastline which can be exploited for seaweed farming. Seaweed can grow prolifically along the coasts of Gujarat and Tamil Nadu and on the islands of Andaman and Nicobar and Lakshadweep. Some other potential seaweed beds include Goa, Mumbai, Pulicat, Ratnagiri, Varkala, and Chilka. About 60 species of seaweed are commercially important out of a total of 700 species of marine algae growing in deep water and intertidal areas of the Indian coast. Since the cultivation of seaweed needs minimal technological requirements and capital, its farming provides an alternative occupation for people dwelling in coastal areas. Surveys were carried out by CMFRI, CSMCRI, and other organisations and they concluded that there are vast seaweed resources along the shorelines of South India and Gujarat on the west coast. The production of dry agar is around 110-132 tonnes produced from 880-1100 tonnes of dry *agarophytes* (agar-producing algae) and the production of algin is 360-540 tonnes from 3600 to 5400 tonnes of *alginophytes* (algin-producing algae).

Seaweed cultivation has emerged as a profitable commercial endeavour in a large number of tropical countries with clean and pollution-free intertidal environments. The export market is quite potential due to the diverse use of seaweeds. It was estimated that India has the potential to produce 1 million tonnes of seaweed (dry) which can provide employment to around 2 lakh fishermen with an average annual income of ₹1 lakh per individual. Seaweed farming also leads to the employment of village women and serves as a tool for women's empowerment, making them financially independent and even the bread earners of their families.

Advantages of Adoption of Seaweed Farming

1) The trade opportunities have been developed because of seaweed farming in developing countries like India promoting village communities, small family farms, and especially women. Seaweed farming has helped women to be economically self-sufficient and at the same time has reduced environmental problems like overfishing and climate change. Hence, seaweed farming proved to be a sustainable green business model for the present and upcoming generations.

2) It does not require resources like land, plant protection appliances, fertilizers, seeds, or freshwater which are costly inputs in farming practices. Hence it reduces the cost of cultivation. Among its necessities, carbon dioxide, sunlight, and saline water are easily available from the environment. Only the farmer needs to spend money on the boat.

3) As Seaweed farming is a flexible enterprise, it helps women in developing countries to get an independent income that can support their lifestyle. An example of women empowerment can be seen in Tanzania where women have emerged as leaders in the seaweed business.

4) Over the years, the sustainability of the global aquaculture system has reduced due to problems like overfishing. Overfishing has led to the decline of species of fish resulting in a reduction in nutritional security. The emergence of seaweed farming in coastal areas can serve as a profitable alternative to fishing and thus can slow down overfishing.

5) Due to climate change, the temperature of the ocean and the concentration of carbon dioxide in seawater are increasing at a rapid rate. Seaweeds being the trees of the sea have the potential to absorb carbon dioxide from seawater and reduce the effects of climate change in oceans.

6) Seaweed farming can help keep the sea coast cleaner by absorbing toxic chemicals such as nitrogen from sewage and agricultural runoff.

Integration of Seaweed Farming with Other Enterprises

Seaweed farming can be well integrated with other allied enterprises to increase the sustainability of the system. Seaweeds are generally seen to incorporate with fish farming to increase the net income from the system. A common example is integrating red seaweed *Kappaphycus* with Cobia fish. *Gracilaria verrucosa* seaweed is integrated with shrimp culture in a brackish water pond. This system has a profound effect on stress reduction on shrimp by absorbing nitrogenous waste from the system and thus also increasing the growth of seaweed. Another type of diversified system includes the cultivation of fed species like fish or shrimp along with mussels and oysters which serve as filtering the particulate organic matter (POM) and seaweeds as dissolved inorganic matter extracting (DIM) species. These three units help to maintain the sustainability of the ecosystem by reducing pollution and also providing a handsome income to the farmer.

Apart from this, seaweed can be cultivated in an integrated system with agricultural crops and livestock. The harvested seaweeds after processing can be used as biostimulants to increase the growth and quality of agricultural crops. Fresh seaweeds can also be fed to farm animals which will increase the body weight of poultry and cattle, increase egg production, raise calcium and iron content in the milk, and reduce methane emissions by ruminants. Seaweeds can also be used as raw materials for making manure. Apart from this, the seaweed also maintains the quality of water by reducing eutrophication caused by the runoff of fertilisers and plant nutrients from the field.

Nutritional Composition of Seaweeds

Seaweeds are very rich in essential elements and generally contain all sorts of vitamins like retinol, thiamine, riboflavin, niacin, pantothenic acid, cobalamin, and folic acid. Among the minerals, potassium, sodium, phosphorus and calcium are also present. They also possess a well-balanced amino acid content. More than 54 trace elements are present, in amounts more than vegetables and plants, helping in the physiological functioning of human beings. As these essential elements and minerals are present in chelated form, they are bioavailable.

Applications of Seaweed in Various Sectors

Seaweeds are widely used in a variety of fields such as food, feed, cosmetics, fertilizer, and textile industries. As a basic utility seaweeds are used as food, livestock feed, biofertilizers for plants, and biofuels. In industry, they are harnessed to produce nutraceuticals, cosmetics, and industrial chemicals.

Besides these, seaweeds also perform some ecosystem services like nutrient scrubbers, carbon dioxide scrubbers, oxygenators, etc.

Seaweeds are consumed as food in countries like China, Japan, Korea and recently in the USA and Europe. The highest consumption of seaweed is found in Korea. Seaweeds are popularly known as "sea vegetables" in many parts of the world. They are rich in protein, vitamins, minerals, omega-3s, calcium, and iron and hence referred to as the medicinal food of the 21st century.

Seaweeds are also used as livestock and poultry feed. They are consumed either as freshly harvested material or as seaweed sap. Seaweeds are known to increase the body weight of poultry and cattle. They help in maintaining the health of the digestive tract, increment egg production, and increase calcium and iron content in milk. They help in better retention of calcium, thus protecting the animal from milk fever. Seaweeds are known to reduce the emission of methane from ruminants.

Seaweed extracts have a stimulatory effect on agricultural crops. The application of seaweed extract increases the germination percentage and seedling vigour of plants. Growth parameters and yield of crops are known to increase by applying different concentrations of seaweed extract. Apart from yield, they are also known to increase the quality of the produce. Seaweed extract also acts as a soil conditioner and activates soil microbial biomass, making the plants resistant to various pests and diseases. Hence seaweed application promotes the sustainable cultivation of agricultural crops.

Carrageenan is a soluble fibre extracted from red seaweeds which helps in stabilizing foods in an organic system. Being a natural food additive, it replaces sugar and salt. It is also used in the production of vegan and dietary supplements which have many health benefits. Seaweeds are used in the production of agar which is used as a replacement for gelatin. It has numerous uses in the food industry like soup thickeners, fruit preservations, in ice creams, and desserts. Seaweed is used in the pharmaceutical industry in the production of Algin which is used in making medicines. Sodium alginate is also produced from seaweeds.

Seaweeds are often used in cosmetic products. Phycocyanin is present in red seaweeds and is used in cosmetic products such as lipsticks and eyeliner. Seaweeds are exploited as cleansing and detoxifying agent. They help in increasing hydration and trans-epidermal water loss from the skin. Seaweeds are known to have anti-aging properties. Hydrocolloids are made from seaweeds which serve as a thickening and gelling agent. They are used in soups, sauces, jam, jelly, etc.

Suitable Site for Seaweed Cultivation

A suitable site for seaweed cultivation is an environment where the annual mean growth rate is more than or equal to 4% per day. Seaweed is generally cultivated in shallow waters of coastal areas having stable seawater and a salinity of not less than 30 ppt. A sandy or rocky bottom with transparent water is also suitable. The optimum temperature is 26-30 and the minimum depth should be 1 m deep during low tides. Areas having mild water currents are generally preferred.

Suitable Cultivable Species in India

Some of the cultivated seaweed species in India are *Kappaphycus alvarezii, Gracilaria edulis, Turbinaria* sp., *Sargassum wightii, Gelidiella acerosa, Gracilaria dura.* For agar production seaweeds, *Gracillaria edulis*, and *Gelidiella acerosa* are used. For alginate production, *Sargassum wightii* and *Turbinaria* are used. *Kappaphycus alvarezii* is most widely used for the production of organic fertilizers for applications in agricultural crops.

Different Methods of Seaweed Farming in India

1) **Single Rope Floating Raft (SRFR) Method:** This is one of the widely used methods of seaweed farming developed by the Central Salt and Marine Chemicals Research Institute, Bhavnagar, Gujarat. Two wooden stakes along with 2 anchor cables are attached to a long polypropylene rope measuring 10 mm in diameter. Generally, the length of the cable taken is double the depth of the water column. The rafts are kept in a floating condition with the help of 25-30 floats. The rope where cultivation will take place is attached to the floating rope. At the lower end of the cultivation rope, a stone is attached to keep it in proper vertical proportion. 10 fragments of the cultivated seaweed are inserted into each rope. The rafts are kept at a distance of 2 m from each other.

2) **Tube Net Method:** A tube net measuring 3 m in length, and 1 m in perimeter with a 25 mm mesh size is used for culturing seaweeds. The tube net is generally attached to a raft made of bamboo. In a raft of 15 m × 3 m around 30 tubes of net is attached at a spacing of 50 cm. The raft is kept floating with the help of 50 litres water cans. A seed rate of 25 kg per unit is suitable.

3) **Monocline Method:** A structure is made with 4 bamboo poles measuring 3 m in length and the four sides are tied with the help of 6 mm rope. Polypropylene rope used for seeding is attached to this structure. 40 seedlings can be planted in one rope at a spacing of 15 cm. The total requirement of seeds is 60-80 kg. Fishing nets are used to prevent grazing by sea herbivores. PET bottles are tied to ropes to increase floating.

Integrated Multi-Trophic Aquaculture (IMTA)

Integrated Multi-Trophic Aquaculture is defined as the aquaculture system integration of different enterprises of commercially important aquatic species whose feeding habits differ from each other. Instead of cultivating a single species, IMTA focuses on imitating a natural ecosystem with the help of a combination of various species. Like the Integrated farming system, the waste of one enterprise acts as the resource for the other in IMTA. In IMTA aquatic species of varying trophic levels are used and wastes or co-products of one cultured species may be used as inputs for the other species. This system reduces the application of inputs and thus increases the sustainability of the production system. The output of the animal component (fish/snails) of IMTA which contains nitrogen and phosphorous is used by the seaweeds for their growth and development. By absorbing the nutrients, the seaweed solves the problem of eutrophication, algal bloom, and finally, Biological Oxygen Demand which results in the deaths of aquatic life. IMTA includes three units of aquatic species in the right proportions – Fed species (Fish or Shrimp) Particulate organic matter filtering species (Mussels, oysters, scallops etc.) and Dissolved inorganic matter extractive species (Seaweeds).

The fed species release POM into the water due to unconsumed food and faeces production. The fed species also produce DIM as nitrate, phosphate and carbon dioxide due to metabolism. The molluscs act as a POM filter, absorbing the POM and enhancing their diet. The seaweeds take up the DIM and use them for growth and development. In this way, the waste of one unit acts as a resource for other units, ultimately reducing pollution and increasing the net return.

A suitable example of IMTA is the integration of red seaweed (*Kappaphycus alvarezii*) with cage farming of Cobia fish (*Rachycentron canadum*) on the coast of Tamil Nadu by a fisherman community under the supervision of ICAR-CMFRI, Mandapam. About 16 rafts made from bamboo measuring 12 feet × 12 feet were constructed and 75 kg of *Kappaphycus* sp. per raft was integrated with one cage. The seaweeds were harvested at 45 days. The productivity of the raft integrated with the cobia cage was 260 kg of seaweed per raft while it was 150 kg of seaweed per raft for the non-integrated one. Hence, an additional yield of 110 kg of seaweed per raft was obtained as a result of integration with the cage. Also, the number of newly emerged apical portions of harvested seaweed was higher in the raft that was integrated. Therefore, it was observed that by integrating *Kappaphycus* and *Cobia* cultures, additional revenue was generated due to the increment of the yield of both species.

Care, Maintenance, and Harvesting

The growth and development of the seaweeds need to be checked two times a week. Their length and biomass should be measured. Environmental and hydrological parameters including pH, turbidity, salinity, water current, temperature, light intensity epiphytes, epifauna, and other predators should be monitored. Netting should be used if necessary to remove sea herbivores. Barnacles and other undesirable algae should be removed through proper cleaning. Tighten any loose ropes. If disease infestation is noted, go for recommended control measures.

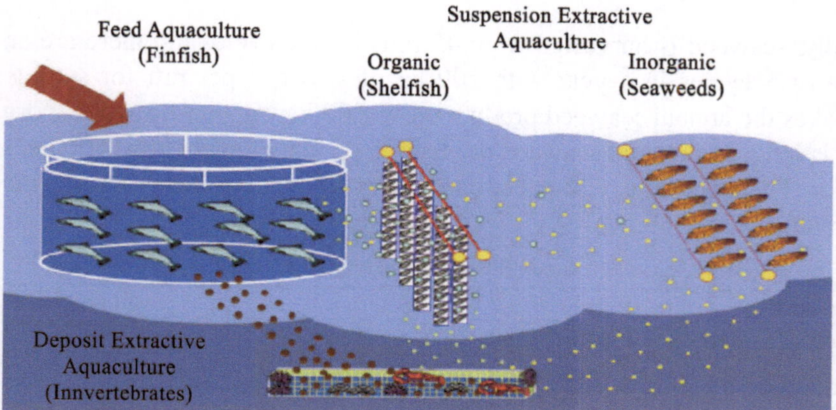

Fig. 1: Diagrammatic representation of Integrated Multi-Trophic Aquaculture (IMTA) [*Source*: Seafood alliance, 2010]

The crop growth period is generally 45 days. The seaweeds are either handpicked or harvested with a knife or scissors. After harvesting, the seaweeds are thoroughly cleaned to remove undesirable particles and dirt. The seaweeds are then sundried for 2-3 days to fetch a good price. Storing the harvested seaweeds should be done in a dry, cool, and well-ventilated place.

Economics of Seaweed Farming

In the case of raft farming of seaweed, the total cost incurred for producing one raft out of bamboo is around ₹1600/-, where fixed costs of production are

₹1000/- and operating cost is ₹600/-. The fixed cost includes the cost of 3-4” diameter hollow bamboo of 12’× 12’ for making the main frame and 4’× 4’ for diagonals of the raft, the cost of a five-toothed iron anchor to hold the rafts, the cost of PP twisted rope for seaweed plantation, the cost of HDPE (High-density polyethylene) braider pieces, the cost of raft framing rope, anchoring rope, the cost of HDPE fishing net to protect the raft bottom, etc. The operating cost includes the cost of seed material, labour, transportation costs and some miscellaneous expenses.

Since the duration of each crop is 45 days, around 4-6 crops can be harvested annually, based on the prevailing climatic conditions. About 150 g of seaweed as planting material has the potential to grow up to 500 g to 1 kg in a period of 45 days. The average yield obtained from a raft of 12’× 12’ size is 260 kg. Out of this total production, 60 kg is kept as seed for the upcoming crop, and the rest is sold. The seaweed is sold on a fresh weight basis or dry weight basis. The farmers earn ₹5 to ₹10 per kg for fresh seaweed and ₹40 to ₹45 per kg of dry seaweed. It was seen that a farmer can earn around ₹10,000 to ₹15000 per month.

Generally, seaweed is harvested from 45 rafts in one cycle, and there are an average of 5 cycles in a year with 200 kg of seaweed per raft for selling. This makes the annual seaweed production 45,000 kg. Again, the average dry weight is 10%, so the total seaweed production on a dry weight basis becomes 4,500 kg. The price of dry seaweed is ₹45 per kg making the gross income ₹2,02,500/-.

Table 1: Analysis of the economic feasibility of seaweed farming from 45 rafts per person

Description	**Amount**
Fixed cost of production of one bamboo raft	₹1,000
Operating cost of production of one bamboo raft	₹600
The total cost of production of one bamboo raft	₹1,600
The total cost of production of 45 bamboo rafts	₹72,000
Gross income of seaweed production from 45 rafts	₹2,02,500
Net income of seaweed production from 45 rafts	₹1,30,500
Benefit-cost ratio (B:C)	2.8

Source: Johnson *et al.*, 2020]

Environmental Benefits from Seaweeds

Generally, seaweeds are valued for their diversified products but they also render ecosystem services. Seaweeds are known to play a crucial role in reducing the adverse effects of climate change. They increase the biodiversity of the sea by giving shelter to a variety of organisms. By absorbing carbon dioxide, they lower the global warming. Being good nutrient scrubbers, they take up dissolved nitrogen, phosphorous, and carbon from the water. They even control organic pollution by absorbing heavy metals from seawater. Seaweeds are also used as oxygenators to increase the oxygen content in the aquaculture system. By taking up the nutrients from water, they prevent the problems of eutrophication.

SWOT Analysis of Seaweed Cultivation in India

The strengths include the availability of a long coastline and vast wasteland belts along the shoreline. It is a low-cost technology with low labour costs and good returns. The availability of resources is also optimal. It provides a viable alternative occupation among the coastal community.

The weakness includes poor automation in seaweed farming and processing and a lack of awareness among people regarding its farming and uses. The research and development section are weak and it also lacks market predictions and technology forecasting.

The opportunities include the chance of an increase in exports and a reduction in imports of seaweeds. Multiple products are being made from seaweed. Seaweed extract has the potential to bring revolution in organic agriculture. It also has good scope for rural entrepreneurship. Among the threats, climate change and global warming are crucial. The lack of preventive measures for diseases and overgrazing is another problem.

Challenges in Seaweed Production in India

1) Due to a lack of diversification, seaweeds are mainly exploited in the production of phycocolloids which alone do not give enough profit that could overcome the high cost of pond production of seaweeds. Seaweed farmers are earning very small amounts from their produce due to selling to cottage-level industries. Hence, harvested seaweeds need to be used for diversified uses like bio-stimulants, pharmaceuticals, biofuel, etc.

2) The occurrence of cyclones during the Southwest and Northeast monsoon has become frequent leading to high turbulence in seawater along with fluctuations in high tides. Hence, the seasonal dependency on seaweed farming is becoming detrimental.

3) Due to the exposure of vast Indian shorelines to the open sea, the frequency of high tides is higher. More research needs to be done for

developing a culture system that would have the potential to tolerate high water dynamics.

4) Among the cultivated species of seaweed, *Kappaphycus* seems to dominate the production system. But there is a scarcity in the availability of quality seed material in both time and quantity in the coastal areas of the country.

5) Lack of proper extension services is a crucial problem for seaweed cultivation. Training and workshops related to seaweed cultivation are very few for educating the farmer, thus it becomes hard to train them.

6) Lack of a well-organised marketing structure is a problem. The involvement of middlemen makes marketing troublesome and thus has small profit in the hands of farmers.

7) Cultivation constraints include grazing on the tips of branches by sea herbivores, attachments of undesirable weeds to the cultured species, and the occurrence of diseases due to high temperature and low salinity problems.

Success Story of Seaweed Farming in India

Mohammed Mariliya, a female seaweed farmer who lives in Munaikadu of Tamil Nadu claims that she has been involved in seaweed farming for ten years and adopted the bamboo raft method of seaweed cultivation. She earns around ₹1000 to ₹1200 per bamboo raft per cycle of cultivation. Her husband Mr. Ajmal Khan used to be a fisherman, but he couldn't earn enough for their well-being. However, after they started seaweed farming, the well-being of the family improved due to better income. During the harsh time of Covid-19 and the ban on fishing, seaweed farming provided them with continuous income from which they could earn their slices of bread and continue the study of their children. This motivated them to keep doing seaweed farming.

Another example is from Puthukudi of Tamil Nadu where seaweed farmer Rajlakshmi says that her husband is a fisherman but he couldn't catch much fish and thus earned less. Earlier she used to be a housewife. When she came to know about seaweed farming and she along with 40 others participated in a 3 days training programme conducted by CMFRI, Mandapam. There she got all the inputs for bamboo raft seaweed farming from CMFRI and started seaweed farming. This has led to an increment in their earnings, she also claims that one could earn more from seaweed farming compared to fishing.

Future Perspective

More research should be carried out on seaweeds to increase their applicability in more sectors. A good infrastructure needs to be set up for the continuous availability of good-quality planting material. The vigour of the planting material needs to be improved through tissue culture methods for getting higher yields and robust growth along with resistance to climate change. More government schemes should be introduced for mass-scale increment in seaweed production. The potential seaweed beds should be identified to increase production. Seaweed site mapping and a pilot study should be initiated. The fisheries department should create deep seaweed farming infrastructure to allow the farmers to move beyond the shallow waters. The seaweed industry lacks a well-organised market infrastructure for better trading of the produce. More and more training programmes and workshops need to be conducted to make the fishermen aware and educate them about seaweed farming as an important sustainable farming practice. Farmers should be encouraged to use seaweeds in poultry, cattle, and fish diets to enhance production. Even our honourable Prime minister says "*Scientific intervention and technologies hold the key to improving productivity in Indian agriculture... We now need to focus on a Blue Revolution... fisheries... ornamental fishes and seaweeds... We need greater research and promotion of coastal seaweeds... Coastal seaweeds have great potential for human health care and agriculture... We should work on scientific methods of seaweed agriculture. Seaweeds are important raw materials... and can play a significant role in improving crop productivity...*"

Conclusion

Seaweed farming and its integration with various allied enterprises will bring revolution to the field of agriculture. Seaweeds are utilized to produce diversified products in various industries. Seaweed farming does not require many costly inputs like farming of land-based enterprises but the revenue generated by them is good and sustainable. Seaweed farming is emerging as a new occupation among the fisherman community which can reduce problems like overfishing. Seaweed farming has proven to be a sustainable green business model for the present and upcoming generations. It also engages women in society in farming activity and emerges as bread earners of families. Seaweeds are consumed as food in countries like China, Japan, and Korea and recently in the USA and Europe. Seaweeds are also used as livestock and poultry feed. Seaweeds are used in the production of agar, algin, phycocolloids, and carrageenan. ITMA focuses on imitating a natural ecosystem with the help of a combination of various species. This system reduces the application of inputs and thus increases the sustainability of the production system. The output of the animal component (fish/ snails) of IMTA which contains nitrogen and phosphorus is used by the seaweeds for their growth and development. By absorbing the nutrients, the seaweed solves the problem of eutrophication, algal bloom, and finally, Biological Oxygen Demand which results in the deaths of aquatic life. There are various constraints and challenges faced by the seaweed industry, which need to be solved by means of research, extension services, and making the farmer aware and educated about the industry. If proper care is taken then there is no doubt in commenting that seaweed farming will emerge as a top revenue generator of the country. Seaweed farming is a labour-intensive process and generates employment at every level, primarily for women. Workers who do the drying, cleaning, sorting, and harvesting could earn approximately ₹500/ day, so, efforts may be prioritized for the formation and promotion of FFPOs, and support to cooperatives and women SHGs in the potential coastal areas of our country.

References

Begum. M., Bordoloi. B.C., Singha. D.D., Ojha. N.J. (2018). Role of seaweed extract on growth, yield and quality of some agricultural crops: A review. Agricultural Research Communication Centre Journal. 39(4): 321-324.

Chopin, T., Troell, M., Reid, G.K., Knowler, D., Robinson, S.M.C., Neori, A., Buschmann, A.H. and Pang, S. (2010). Integrated marine-trophic aquaculture, part 1, Seafood alliance.

Chopin, T. and Robinson, S. (2004). Defining the appropriate regulatory and policy framework for the development of integrated multi-trophic aquaculture practices: introduction to the workshop and positioning of the issue. Bulletin of Aquaculture Association of Canada 104: 4-10.

Dhandapani, R., Elangovan, M.P., Subbarayalu, R., Khalifa, A.Y.Z., Paramasivam, R., Sankaranarayanan, Thangavelu, S. ad Selvakumar, V. (2022). Small, Large-Scale Production and Cost-Benefit Analysis and Marketing of Agar from Gelidium. In: Industrial Microbiology Based Entrepreneurship, Springer, 127-145.

Johnson, B., Tamilmani, G., Sakthivel, M., Anikuttan, K.K., Ramesh Kumar, P., Sankar, M., Jayakumar, R., Priya, K.M., Uma, E.K. and Gopalakrishnan, A. (2022). Seaweed Farming – Success Story: Mohammed Mariliya Munaikadu village Ramanathapuram.

Johnson. B., Abdul Nazar, A.K., Jayakumar, R., Tamilmani, G., Sakthivel, M., Rameshkumar, P. Anikuttan, K.K and Sankar, M. (2020). Seaweed Farming and Integrated Multi-Trophic Aquaculture (IMTA). IMTA Course Manual, CAU Tripura. pp. 82-87.

Nardelli. A.E., Chiozzini. V.G., Braga. E.S. and Chow. F. (2018). Integrated multi-trophic farming system between the green seaweed Ulva lactuca, mussel, and fish: a production and bioremediation solution. Journal of Applied Phycology. 31: 1573-1576.

Poza, S.G., Leandro, A., Cotas, C., Cotas, J., Marques, J.C., Pereira, L. and Goncalves, A.M.M. (2020). The Evolution Rod of Seaweed aquaculture: Cultivation Technologies and the Industry. International Journal of Environmental Research and Public Health. 17: 1-13.

Rimmer, M.A., Larson, S., Lapong, I., Purnomo, A.H., Pong-Masak, P.R., Swanepol, L. and Paul, N.A. (2021). Seaweed Aquaculture in Indonesia Contributes to Social and Economic Aspects of Livelihoods and Community Wellbeing. Sustainability, 13(19): 10946.

Sharma, P.C. (2016). Innovative Saline Agriculture, Springer Science and Business Media LLC, (DOI: 10.1007/978-81-322-2768-7).

Singh, S.B., Kaur, I., Tuteja, M., Gopal, R.N. and Suresh, P.N. (2021). Entrepreneurship Development Through Seaweed Business by Cooperatives, A workshop jointly organized by Department of Fisheries; Ministry of Fisheries, Animal Husbandry and Dairying, GoI; Network for development of agricultural cooperatives in Asia and the Pacific; Laxmanrao Inamdar, National Academy For Cooperative Research and Development, pp. 37-52.

Wang, X., Olsen, L.M., Reitan, K.I. and Olsen, Y. (2012). Discharge of nutrient wastes from salmon farms; environmental effects, and potential for integrated multi-trophic aquaculture. Aquaculture Environment Interaction. 2: 267-283.

13

Entrepreneurship Opportunities in Integrated Farming System Through Efficient Utilisation of Weeds

G. Manisankar[1], M. Banerjee[1], G.C. Malik[1], Sarath Kumar Duvvada[1] and Kajal Sengupta[2]

[1]Department of Agronomy, Palli Siksha Bhavana (Institute of Agriculture) Visva Bharati, Shantiniketan, Bolpur, West Bengal
[2]Division of Agronomy, Faculty of Integrated Rural Development and Management (IRDM), School of Agriculture and Rural Development Ramakrishna Mission Vivekananda Educational and Research Institute [Deemed to-be-University], Narendrapur Campus, Kolkata, West Bengal

Abstract

Weeds are plants that emerge uninvited and at the wrong time. Weeds are the major biotic constraint responsible for declining agricultural productivity globally. Depending on the growing environment, weeds compete with crops for space, nutrients, water and light. At present in India, a shortage of labour during peak agricultural seasons necessitates the utilisation of herbicides to effectively control the majority of weeds. The application of herbicides may reduce labour cost and save time but at the same time it pollutes the environment and leads to the loss of weed wealth. Indeed, weeds also have some economic importance similar to crops. However, the lack of awareness about their potential uses, proper handling and processing knowledge among the farming community hinders their efficient utilisation. Around 75% of weeds and their products have viable economic importance and are used in different industrial sectors including ancient medicine, modern medicine, cosmetic sector and food sector. Hence, efficient utilisation of weeds is a potential and novel option for entrepreneurship. Therefore, proper training and awareness can be provided to students and the farming community to enhance entrepreneurship and develop the rural Indian economy. Efficient utilisation of weeds not only creates entrepreneurship but also helps increase the per capita income of farmers and reduce environmental pollution caused by synthetic herbicides. Undoubtedly, similar to other agricultural and allied sectors, efficient utilisation of weeds also helps increase the national agricultural economy in the future.

Keywords: Weed, Weed utilisation, Weed uses, Entrepreneurship through weeds.

Introduction

From a scientific perspective, weeds are plants that grow in locations and times when we desire to cultivate other plants or no plants at all. Weeds are uninvited, rapidly spreading, competitive and environmentally detrimental vegetation that can be found in both cultivable and non-cultivable lands. Globally, approximately 30,000 plant species have been classified as weeds, out of which 18,000 pose significant threats to crop productivity, leading to substantial losses. Among the different factors causing crop productivity losses, weeds are a major biotic constraint and account for 33%. Weeds can be efficiently controlled through various methods, including cultural practices (tillage, appropriate planting methods, selecting suitable crop varieties, adjusting planting density, implementing proper irrigation and fertiliser application, ensuring adequate drainage and adopting appropriate cropping systems), physical methods (hand weeding, hand hoeing, digging, mowing, cutting, dredging, chaining, burning and flaming), chemical methods involving the use of herbicides and biological approaches.

Among the different agricultural operations, weed management alone accounts for 20 percent of the total cost of cultivation. In recent times, the rising agricultural labour wages and shortage of labour during critical agricultural operations have prompted the adoption of herbicidal weed management. This herbicidal weed management has reduced 66% of the cost incurred in manual weeding (Chandrakanth, 2022). However, the application of herbicides may reduce the labour costs and save time, but at the same time, it pollutes the environment and leads to the loss of weed wealth. On the other hand, the eradication of weeds in the farming land may enhance crop productivity but efficient utilisation of weeds may create a new entrepreneurship opportunity.

Among the different sectors contributing to the Indian economy, agriculture is a notable one. In recent days, several new ideologies and advanced technologies have been developed to enhance crop productivity. These technologies have not only enhanced agricultural productivity but also opened up employment opportunities. Among the different sectors of occupation, the agricultural sector is considered a peaceful sector by numerous people. Because of that, now a days we hearing news about techies quitting IT jobs and doing agriculture. Hence, from people's point of view, agriculture is a peaceful sector for employment and economic generation at present.

Income generation through the agricultural sector does not only depend on the cultivation of crops and marketing the final product. It entirely depends on the efficient utilisation of all available farm resources. Agricultural GDP is not solely dependent on agricultural productivity; it also relies on the economy of allied sectors like beekeeping, sericulture and livestock farming. Efficient utilisation of all these resources is essential to enhance agricultural GDP. In recent days, several entrepreneurship opportunities have emerged from the agricultural sector. Among the different opportunities, the least noticed and less utilised option is the efficient utilisation of weeds.

According to an Ethno-Botanical Survey, numerous food plants that have been traditionally consumed are often overlooked by scientific researchers and agriculturists in modern-day crop production, as they are categorised and treated as weeds. However, efficient utilisation of these unwanted plants is a novel and notable entrepreneurship opportunity in the agriculture sector to enhance the economy. Obviously, these unwanted plants from every cropping system have a viable economic importance. Due to a lack of proper education and awareness among farmers about the potentiality of weeds, till today these plants are not efficiently utilised. According to Ramachandra Prasad (2011), more than 75 per cent of the weeds possess medicinal properties and due to the lack of handling knowledge and marketing opportunities, their commercial utilisation remains limited. These weeds hold tremendous potential for medicinal purposes, but their untapped value stems from the challenges associated with effectively harnessing their benefits for commercial purposes.

Weeds have enormous potential for commercial utilisation. Weeds can be efficiently utilised for culinary purposes, as fodder for animals, as medicinal plants, aromatic plants, aesthetic plants, to prevent soil erosion, for soil reclamation and as a nematicide. At the same time progressive awareness and training among the farming communities regarding the entrepreneur opportunities from weeds are lacking. Keeping these points in view, this book chapter aims to describe the economic importance and entrepreneurship opportunities through efficient utilisation of weeds.

Importance of Weeds in Indian Economy

India is predominantly an agrarian country, with agriculture playing a crucial role in its economy and contributing around 25 percent to the gross domestic product (GDP). A significant portion of the population, approximately 70 per cent, resides in rural areas and depends on agriculture and its allied sectors for their livelihoods. By focusing on the development of the Indian economy through agriculture and allied sectors, significant strides can be made in uplifting the rural community and promoting overall economic growth. In recent days, several entrepreneurship opportunities have increased in the agricultural sector to enhance the agricultural GDP. More than that, the agricultural sector not only provides food to the people, but also serves as an input for a versatile industrial sector. Among the different crop production constraints in India, weeds alone cause crop production losses of 1100 crores to Indian agriculture (Gharde *et al.*, 2018). On the other hand, 75% of weeds have potential economic viability. Hence, efficient utilisation of weeds in an economical way may be a viable option to overcome crop production losses and protect the soil environment from synthetic herbicides.

To enhance weed utilization and develop new entrepreneurship opportunities, proper awareness and training programs are needed regarding their economic importance and processing methodologies. When entrepreneurial opportunity arises, the first question that comes to mind is how to market the commodities. Currently, there are various e-commerce platforms available in India for buying and selling the agricultural commodities that includes Flipkart, Amazon, Bighaat, Kisan Mandi, Krishi Market etc. Selling of products on e-commerce platforms offers sellers convenient price fixation and also improve the seller-consumer relation.

Economic Importance and Entrepreneurship Opportunities in Weeds

This title pertains to the economic importance and entrepreneurship opportunities of important weeds that commonly spread over the different agro-climatological zones in India.

1. *Acalypha indica* (L.)

Common name: Indian Acalypha

Scientific name: *Acalypha indica* (L.)

Family : Euphorbiaceae

Economic part: Entire plant

Potential Uses: *Acalypha indica,* also known as Indian nettle or Acalypha, possesses medicinal properties and is utilised for various purposes. The leaf decoction of *Acalypha indica* is commonly used to alleviate earaches and infections when applied directly to the ears. Additionally, a paste made from powdered leaves is applied topically to treat skin infections. A decoction created by combining *Acalypha* leaf powder and garlic is administered as a remedy for intestinal worms. Furthermore, a decoction prepared from dried leaves of *Acalypha* is utilised as an aphrodisiac and often employed as a massage cream to alleviate joint pain. Leaf powder obtained from *Acalypha* is used as an antiseptic to enhance wound healing. The crushed leaf juice of *Acalypha*, when mixed with salt, is applied to treat scabies and other skin ailments. The seed powder of *Acalypha* is used for amoebiasis, a condition caused by amoebic parasites. The fruit pulp of *Acalypha* is believed to possess properties beneficial for managing diabetes. In Siddha medicine, *Acalypha* is highly regarded and believed to have rejuvenating effects on the body (Ramachandran, 2008). The polyphenols and steroids present in *Acalypha* help to lower blood sugar levels and play a significant role in controlling diabetes. In the cosmetic sector, the leaves of *Acalypha* are used to remove unwanted hair. To ensure this, apply *Acalypha* leaf paste on the skin after it has been exfoliated with warm water. Due to its wide range of applications and medicinal properties, *Acalypha indica* is included in the Pharmacopoeia of India. It is specifically listed as an expectorant, used for treating conditions such as asthma and pneumonia.

Entrepreneurship Opportunity: *Acalypha indica* has potential benefits in the cosmetic industry. Ancient medicine believed that it helps rejuvenate the body. Hence, manufacturing *Acalypha indica* soap and serum creates a new entrepreneurship opportunity in the cosmetic and beauty sector.

2. *Amaranthus viridis* (L.)

Common name: Green amaranth

Scientific name: *Amaranthus viridis* (L.)

Family : Amaranthaceae

Economic part: Whole plant

Potential Uses: Young plants and leaves of *Amaranthus* viridis are occasionally used as green leaf vegetable. Amaranthus is also a very good livestock fodder except piggery and good green manure crop. In Nigeria, entire plant is used as a blood cleanser and base root is used to treat diarrhoea. It is one of the typical vegetables used in Odia cuisine Saaga. Green amaranthus seeds are used as an additive in biscuits and snacks making.

Amaranthus seed extract is an excellent skin moisturiser. In India and Nepal, it is commonly utilised for reducing labour pain and has been recognised for its antipyretic properties, which help to lower fever (Kiritikar and Basu, 1987).

Entrepreneurship Opportunity: Amaranthus is considered a rich source of protein. It contains the essential amino acid lysine which is present in human hair but the body cannot produce it on its own. Therefore, the powder or liquid shampoo made from amaranthus acts as a natural hair strengthener. Amaranthus also contains high levels of squalene (an organic compound), omega-6 and omega-9 fatty acids. The leaf extract of amaranthus is quickly absorbed by the skin and acts as a natural skin moisturiser.

3. *Avena fatua* (L.)

Common name: Green amaranth

Scientific name: *Avena fatua* (L.)

Family : Poaceae

Economic part: Seed

Potential Uses: The seed extract derived from wild oats is used as an emollient for the skin. It works very well to treat hot, dry, irritated and inflamed skin. Additionally, wild oats help in restoring, rejuvenating and strengthening nerve cells.

Entrepreneurship Opportunity: Wheat is the most cultivated cereal crop after rice in India. Wild oat is an associated weed of wheat. Extracting of milk from wild oat seeds presents a fresh entrepreneurship opportunity in the cosmetic industry as it helps moisturise the skin.

4. *Cardiospermum halicacabum* (L.)

Common name: Balloon vine or Mudakathan keerai

Scientific name: *Cardiospermum halicacabum* (L.)

Family: Sapindaceae

Economic part: Whole plant

Potential Uses: Balloon vine is also called "*Love in a Puff*" and is one of the top 10 sacred flowers of Kerala. A 100-gram serving of balloon vine contains approximately 61 calories including 4.7 grams of protein, 0.6 grams of fat and

9 grams of carbohydrates. In addition to its macronutrient content, balloon vine is rich in minerals such as calcium, iron and phosphorus. Balloon vine has antibacterial, antifungal, antiparasitic, antidiarrheal, anxiolytic, rubefacient, antipyretic, antiinflammatory, anticonvulsant and anticarcinogenic medicinal properties. A leaf extracted solution of balloon vine is used to enhance hair growth, reduce scalp issues and cure earaches. The oil derived from balloon vine is a potent remedy for rheumatism, joint pain and body pain. The leaves of *Cardiospermum* are used to make herbal tea (Shaily and Sarmad, 2018). The extract obtained by boiling the leaves of balloon vine in water has mild anti-ulcer properties. In Bangladesh, pills made from balloon vine paste are commonly used as a treatment for asthma, providing potential relief for respiratory symptoms. Additionally, the leaf extract of balloon vine is recognised for its potent mosquito repellent properties. Importantly, balloon vine has the ability to enhance sperm count and sperm motility in humans. Regular intake of balloon vine for a month helps to significantly increase sperm count and serum testosterone levels.

Entrepreneurship Opportunity: Balloon vine paste is a popular remedy for joint pain and shoulder pain. In addition to that, fresh green leaves of balloon vine (Mudakathan) are ground into a liquid and mixed with dosa flour to make Mudakathan dosa. Hence, marketing Mudakathan leaf powder is a great entrepreneurial option. The fruits of *Cardiospermum halicacabum* have numerous health benefits for men. Marketing freshly collected *Cardiospermum* fruits is also a viable entrepreneurial option.

5. *Chenopodium album* (L.)

Common name: Lamb's quarters/Pigweed/Bathua

Scientific name: *Chenopodium album* (L.)

Family: Amaranthaceae

Economic part: Leaves

Potential Uses: Leaves and young shoots of *Chenopodium album* are consumed as a leafy green vegetable (Benoliel, 2011). In North India, young shoots and leaves of this plant are used to prepare dishes like soups, curries and paratha stuffed breads. *Chenopodium album* leaves have enormous medicinal uses; They help treat kidney stones, swelling, anaemia, heart disease and jaundice. The leaf juice of *Chenopodium album* is used as an ancient medicine to treat intestinal parasites and increase blood haemoglobin levels. Also, this leaf extract is considered a heart and liver tonic (Anupama, 2015; Saini and Saini, 2020).

Entrepreneurship Opportunity

Chenopodium leaves and seeds have innumerable health benefits like any other leafy vegetables. During the winter season, *Chenopodium album* leaves may be marketed as a green leaf vegetable.

6. *Cynodon dactylon* (L.)

Common name: Bermuda grass

Scientific name: *Cynodon dactylon* (L.)

Family: Poaceae

Economic part: Entire plant

Potential Uses: *Cynodon dactylon* holds great religious significance in India, being considered one of the most sacred plants after tulsi (holy basil). The leaf extract of *Cynodon dactylon* is known for its astringent properties and is commonly used to control or stop bleeding from cuts and wounds. (Chopra *et al.*, 1999). According to the Ayurvedic system of medicine, *Cynodon dactylon* is considered an appetiser, anthelmintic (used to expel parasitic worms), antipyretic (fever-reducing) and alexiteric (immunity-boosting) agent. In the Unani system of medicine, Bermuda grass is used as a brain and heart tonic, indicating its potential benefits for cognitive function and cardiovascular health. In the homeopathic system of medicine, it is employed to address a wide range of skin problems and bleeding conditions (Nagori and Solanki, 2011). According to Melillo (1987), a finely powdered mixture of *Cynodon* is used in asthma inhalers. This grass is mostly preferred for landscaping because of its fast growing and wide-spreading nature. *Cynodon dactylon* roots were used as a substitute for sarsaparilla (Vetiver) and the decoction obtained from *Cynodon* is used as a diuretic substrate. Consuming *Cynodon* juice along with honey two to three times a day is effective in treating menorrhagia. The application of *Cynodon* leaf paste in the lower abdomen helps to stop vaginal bleeding. A decoction obtained from *Cynodon* mixed with sugar is used to cure urine retention problems. According to Saikia *et al.* (2006), Ayurvedic physicians often recommend the use of *Cynodon dactylon* for treating specific gynaecological conditions. It is commonly used for managing menorrhagia, which refers to excessive bleeding during the menstrual cycle. Additionally, *Cynodon* is believed to be beneficial for regulating irregular menstrual cycles and addressing habitual abortion (repeated miscarriage). Furthermore, this herb is known for its potential in combating polycystic ovarian syndrome (PCOS), a hormonal disorder affecting the ovaries. The plant extract derived from *Cynodon* is very effective against snakebite therapy and anti-snake venom (Selvanayagam *et al*., 1995). The fibre extracted from Bermuda grass is used in manufacturing the agarbatti sticks.

Entrepreneurship Opportunity: The leaf juice of *Cynodon* has immense health benefits. It is one of the most preferred juices in the early morning to cleanse the stomach and aid digestion. However, it is not readily available in most of the developed cities. Hence, manufacturing packaged *Cynodon* juice will surely open new entrepreneurship market. Cultivation of *Cynodon* grass and selling it in the devotional places is the viable option for day to day earnings. It is one of the best-selling and most demanded grasses during Aadi Perukku (*Adi* monsoon festival), Ganesh Chaturthi festival and many other festivals.

7. *Cyperus rotundus* (L.)

Common name: Purple nutsedge

Scientific name: *Cyperus rotundus* (L.)

Family: Cyperaceae

Economic part: Tubers and Leaves

Potential Uses: *Cyperus* is one of the carbohydrate-enriched foods consumed by modern hunter-gatherers. It was consumed by humans during the Mesolithic period, Neolithic period and the Rajasthan famine (Bhandari, 1974). Ancient and modern Ayurvedic medicine has observed that the *Cyperus* plant is effective in treating fever, digestive disorders, dysmenorrhea, inflammation and muscle relaxation. The leaf and tuber extract of *Cyperus* contains a high concentration of auxin and phenols, acting as a plant growth regulator to promote root cuttings (Yamashita *et al.*, 2017). The antibacterial qualities of *Cyperus* nuts were utilised to prevent tooth disease in Sudanese citizens 2000 years ago (Traci Watson, 2014). The chemical compounds sesquiterpene and rotundone derived from *Cyperu*s are known for their aromatic properties. These compounds are utilised to enhance the spicy aroma in black pepper and contribute to the distinctive peppery taste found in Australian Shiraz wines (Siebert *et al.*, 2008). The extracted tuber solution from *Cyperus* inhibits the cariogenic activity of *Streptococcus mutans* (Yu *et al.*, 2007). According to Soltan and Zaki (2009), the hydroalcoholic extract of *Cyperus* used for its anti-viral activity against HSV (Herpes simplex-1 virus), POLIO (Poliomyelitis-1 virus) and VSV (Vesicular stomatitis virus). Essential oils extracted from *Cyperus rotundus* exhibit good anti-candida activity (Duarte *et al.*, 2005). Dried *Cyperus* plants were used to make basketry, caps and mats (Simpson and Inglis, 2001). Tuber extract and fine tuber powder are used as additives in perfumes, talcum powder and skin conditioners in the cosmetic industry. The tuber solution obtained from *Cyperus* is used as an insect repellent.

Entrepreneurship Opportunity: Extraction of auxin from *Cyperus rotundus* plant helps prepare and market the organic plant growth regulator. Most of the perfumes available in the market are synthetic. Therefore, tuber extract and tuber powder obtained from the *Cyperus* plant open up opportunities for organic perfume entrepreneurship.

8. *Echinochloa colona* (L.)

Common name: Jungali Rice

Scientific name: *Echinochloa colona* (L.)

Family: Poaceae

Economic part: Entire plant

Potential Uses: *Echinochloa colona* commonly known as jungle rice or wild barnyard grass, is highly regarded as valuable fodder for various types of livestock Form (Quattrocchi, 2006). It is also considered a famine food in Sudan. The seed flour of jungle rice is used to prepare porridge or bread (Lansdown, 2013). According to Devi *et al.* (2017), *Echinochloa* is used to solve spleen and haemorrhage problems. The dried stems of *Echinochloa* are used for weaving natural door mats and caps (Kumari and Saggoo, 2015). In India, the seeds of *Echinochloa colona* have culinary applications as well. They are used to prepare a traditional food dish called Khichadi, which is consumed during festival fasting days (Sumitra and Parul, 2018). The seeds of *Echinochloa colona* are boiled in water and used as a substitute for rice in the region of Rajasthan.

Entrepreneurship Opportunity: Dried stems of *Echinochloa colona* can be used to make hand-crafted natural products like door mats and caps. Jungle rice can also help in giving the proper shapes for hand-crafted mud sculptures.

9. *Eclipta prostrata* (L.)

Common name: False Daisy

Scientific name: *Eclipta prostrata* (L.)

Family : Asteraceae

Economic part: Leaves

Potential Uses: *Eclipta* oil is used to prevent hair fall, hair breakage and early greying of hair. Additionally, it supports overall hair growth and strengthens the hair roots. The leaf paste of false daisy is used to treat baldness and is also a very effective liver tonic for controlling liver metabolism (Meenakshi, 2019). False daisy has the potential to be used in manufacturing hair dye particularly in blue and black hair dye production, and the extracted dye is also used for tattooing. In addition to its well-known applications, false daisy aids in blood purification, cures poor eyesight, rejuvenates the body and treats cracked heels. In Korea, the leaf extract of *Eclipta* is traditionally used as an antidote for snake bites and scorpion stings.

Entrepreneurship Opportunity: False daisy has numerous entrepreneurship opportunity. At present, most of the hair care products available in markets are synthetic. However, *Eclipta prostrata* offers overall hair support naturally. Hence, marketing *Eclipta prostrata* shampoo, powder and dye is one of the promising ways to become a new entrepreneur. In addition to that, based on my observation from e-commerce platforms, there is a huge demand from consumers for fresh *Eclipta* plants.

10. *Eichhornia crassipes* (L.)

Common name: Water Hyacinth

Scientific name: *Eichhornia crassipes* (L.)

Family : Pontederiaceae

Economic part: Flower and Leaves

Potential Uses: Botanical solution extracted from water hyacinth is used as a plant nutrient substitute and it contains 2.94% N, 0.94% P and 0.16% K. Water hyacinth leaf fibre is used in sustainable textile technology and is regarded as an organic garment fibre. Fried hyacinth beans are one of the oldest Chinese remedies for improving digestion and have good digestive properties. It is the ideal choice for treating many skin-related issues in the cosmetics sector due to its antimicrobial, antifungal and antibacterial qualities. In the Philippines, the combination of water hyacinth juice and lemon juice is used as a potent anti-inflammatory agent. In Kenya, flowers are used to treat irregular periods in new mothers and boiled water hyacinth is used to encourage lactation. Composting, vermicomposting, biogas production, biofuel production and bioremediation are further uses for water hyacinth.

Entrepreneurship Opportunity: Extraction of water hyacinth fibres for use in organic garments. Collection of water hyacinth from aquatic sources for composting and vermicomposting. It can be used as an organic plant growth supplement.

11. *Mimosa pudica* (L.)

Common name: Touch-me-not

Scientific name: *Mimosa pudica* (L.)

Family : Fabaceae

Economic part: Entire plant

Potential Uses: Crushed and pasted roots of the touch-me-not plant are known to have beneficial properties for treating minor skin ailments such as boils and acne. The application of this paste can lead to overnight healing, providing relief and improving the condition of the affected skin. Interestingly, the roots of *Mimosa pudica* also contain endophytes, which are microorganisms that form a symbiotic relationship with the plant. These endophytes play a role in the plant's defence mechanisms by producing secondary metabolites such as terpenoids, steroids, alkaloids, phenols and quinines. These compounds protect the plant from various pathogens and contribute to its overall health and resilience. Additionally, *Mimosa pudica* is known for its potential use in phytoremediation, which is the process of using plants to clean up or remove pollutants from the environment. According to Ahamed *et al*. (2012), aqueous extracts from dried leaves of *Mimosa pudica* are used in Mexico to alleviate symptoms of depression. This plant is also considered a mood enhancer and is believed to improve blood circulation. In the region of Haryana, *Mimosa pudica* leaves have been traditionally used to increase sexual potency in men. Additionally, in Madhya Pradesh, both the leaves and roots of this plant are utilised for treating kidney diseases and piles. The root paste of *Mimosa pudica* is fried in castor oil and applied to deep cut wounds to help control bleeding. This method is believed to have haemostatic properties and aid in the wound healing process. Additionally, in West Bengal, a decoction made from the roots of *Mimosa pudica* is mixed with water and used as a gargle to alleviate toothache. The alkaloid mimosine present in *Mimosa pudica* is used for potent anti-proliferative and apoptotic effects. The aqueous extract of touch-me-not root has a significant neutralising effect on the venom of the monocled cobra. This extract has the ability to inhibit the myotoxicity (muscle-damaging properties) and enzyme activity of cobra venom. Touch-me-not also has the potential to fix and convert atmospheric nitrogen (ENVIS, 2022).

Entrepreneurship Opportunity: *Mimosa pudica* leaf powder, extracted oil and seeds are some of the best-selling herbal products on e-commerce sites due to their immense benefits. However, proper processing and handling knowledge is essential for manufacturing these products.

12. *Parthenium hysterophorus* (L.)

Common name: Congress grass

Scientific name: *Parthenium hysterophorus* (L.)

Family : Asteraceae

Economic part: Entire plant

Potential Uses: *Parthenium hysterophorus* is indeed used as an additive in cattle manure. Incorporating *Parthenium* at a 10% concentration into cattle

manure can result in increased methane generation (60-70%) during anaerobic digestion processes. Therefore, it is one of the potential substrates for biogas production (Gunaseelan, 1987). Biochar production from *Parthenium hysterophorus* through pyrolysis has been successfully achieved showing potential for addressing negative carbon dioxide emissions (Kumar *et al.*, 2013). According to Biradar and Patil (2001), composting *Parthenium* along with *Eudrilus eugeniae* helps support the growth of worms indicating it is a good substrate for vermicomposting. Due to its abundant lignocellulosic biomass, *Parthenium* serves as a valuable raw material for the production of various types of paper with sufficient strength and quality. The crude extracted from *Parthenium* leaves is used to suppress steel corrosion in acidic conditions (Ji *et al.*, 2012). According to Chanderkala *et al.* (2012), phytophagous insect larvae *Bombyx mori* showed tolerance against pure parthenin toxin. It was also noticed that feeding *Parthenium* along with a diet supplement helps increase cocoon yield and silk production. Another potential benefit of *Parthenium hysterophorus* extract its use in synthesising the zinc oxide and silver nanoparticles with reasonable stability (Rajiv *et al.*, 2013 and Kumar, 2012).

Entrepreneurship Opportunity: *Parthenium hysterophorus* is one of the most noxious and troublesome weeds in India and proper eradication methods have not yet been discovered. However, efficient utilisation of *Parthenium* for composting and vermicomposting can help to enhance organic manure production. This presents a new entrepreneurial opportunity in areas with abundant *Parthenium* populations. With proper research and scientific advice, it may be recommended as a feed for silkworms as an alternative to mulberry.

13. *Phyllanthus niruri* (L.)

Common name: Gale of the wind

Scientific name: *Phyllanthus niruri* (L.)

Family: Phyllanthaceae

Economic part: Entire plant

Potential Uses: *Phyllanthus niruri* is an excellent remedy for removing or reducing the size of stones that form in the urinary tract. The decoction extracted

from *Phyllanthus niruri* is very effective for diabetes and chest pain. The dried plant powder of *Phyllanthus niruri* is mixed with gruel water and taken for ulcer problems. Two to three time a day intake of 10 to 20 ml of *Phyllanthus niruri* juice is effective against jaundice. *Phyllanthus niruri* is used as an ingredient for more than 175 ayurvedic formulations. *Phyllanthus niruri* fruit is commonly used to treat haemorrhages, diarrhoeas, dysentery, jaundice, cough and anaemia. It is also used to manufacture health care and personal products like hair oil, dye, face cream and tooth powder (Rastogi and Mehrotra, 1991).

Entrepreneurship Opportunity: The leaf powder and fruit juice of *Phyllanthus niruri* have vast health benefits. Moreover, the fruit juice of *Phyllanthus niruri* is commercially available on most of the health care e-commerce platforms. However, it is not available in many developed cities. Therefore, manufacturing packaged *Phyllanthus niruri* juice and leaf powder will surely open up a new entrepreneurship market.

14. *Solanum nigrum* (L.)

Common name: Black nightshade

Scientific name: *Solanum nigrum* (L.)

Family : Phyllanthaceae

Economic part: Entire plant

Potential Uses: Black nightshade tender shoots and leaves are regularly consumed as leafy vegetables worldwide. Since ancient times, *Solanum nigrum* has served as a source of food and is considered a Chinese famine plant (Henderson, 1974). Black nightshade ripe berries were used to make delicious jam with wonder berry. In addition to being beneficial in treating liver cirrhosis, freshly prepared *Solanum nigrum* leaf extract also acts as an antidote to opium poisoning. In Pakistan, the powdered form of the aerial parts of black nightshade has been identified as a potential antiulcerogenic agent, helping alleviate ulcers or prevent their formation (Akhtar and

Muhammad, 1989). On the other hand, the immature fruits of *Solanum nigrum* have been found to contain steroidal glycosides in Japan. These compounds have shown significant anticancer activity, indicating their potential for use in cancer treatment or prevention (Saijo *et al.*, 1982).

Entrepreneurship Opportunity: Cultivation of *Solanum nigrum* in farming land or kitchen garden or home garden and selling it as a leafy vegetable in the local market is a viable option for day to day income. Manufacturing of *Solanum nigrum* fruit jam and leaf powder are another novel entrepreneurship option.

15. *Tridex procumbens* (L.)

Common name: Mexican daisy

Scientific name: *Tridax procumbens* (L.)

Family : Asteraceae

Economic part: Leaves

Potential Uses: *Tridax* leaves have therapeutic properties and are used to treat diarrhoea, dysentery and catarrh. According to Martin Quintal (2009), the leaf extract of *Tridax* has been traditionally used as an antiseptic to treat fresh cuts, wounds, burns and anaemia. The extract is applied topically to the affected area for its potential healing and antiseptic properties. Additionally, extracts obtained from *Tridax* have demonstrated antimicrobial activity against both gram-positive and gram-negative bacteria. This suggests that, the plant possesses potential antimicrobial benefits (Gubbiveeranna and Nagaraju, 2016). Synthesising silver and copper nanocomposites are another potential benefit of *Tridax* leaf extract. The liquid derived from *Tridax* is used as a potent mosquito repellent.

Entrepreneurship Opportunity: With proper knowledge about Ayurvedic medicine, *Tridax procumbens* leaf powder and leaf extract may be marketed as an herbal product for wounds and cuts.

Future Prospects

- In the agricultural education system, potential uses of weeds and their entrepreneurship opportunities may be added as a unit in the weed management course to provide knowledge and awareness among students about entrepreneurship prospects associated with weeds.

- Providing proper training programs about handling and processing of weeds to the rural farming community helps to boost the Indian rural economy and rural employment.

Conclusion

From the above discussion, it can be concluded that weeds also have viable economic importance similar to crops and are used as inputs for different industrial sectors. Therefore, proper training and awareness can be provided to students and the farming community to enhance entrepreneurship and develop the rural Indian economy. Efficient utilisation of weeds not only creates entrepreneurship but also helps increase the per capita income of farmers and reduce environmental pollution caused by synthetic herbicides. Undoubtedly, like other agricultural and allied sectors, efficient utilisation of weeds also helps increase the national agricultural economy in the future.

References

Ahmad, H., Sehgal, S., Mishra, A. and Gupta, R. (2012). Mimosa pudica L. (*Laajvanti*): An overview. Pharmacognosy reviews, 6 (12): 115-124.

Akhtar, M.S. and Muhammad, M. (1989). Evaluation of the gastric antiulcerogenic effects of *Solanum* species of economic importance. The Annals of the Missouri Botanical Garden, 75 (3): 8.

Anupama. (2015). Bathua (*Chenopodium album*) Health Benefits. (Unpublished). [https://www.bimbima.com/ayurveda/health-benefits-of-bathua-chenopodium-album/335/]

Benoliel, D. (2011). Northwest Foraging: The Classic Guide to Edible Plants of the Pacific Northwest. Vol. 1. pp. 1-34. Skipstone Publishers, United States of America.

Bhandari, M.M. (1974). Famine Foods in the Rajasthan Desert. Economic Botany, 28 (1): 73-78.

Biradar, A.P. and Patil, M.B. (2001). Studies on utilisation of prominent weeds for vermiculturing. Indian Journal of Weed Science, 33 (3/4): 229-230.

Chanderkala, M.V., Dharani, P. and Patil, R.R. (2012) Investigation on the application of *Parthenium hysterophorus* extracts as feed additives for young larvae of silkworm *Bombyx mori* L. Agricultural Science Research Journals, 2 (8): 449-452.

Chandrakanth, M.G. (2022). Herbicide usage in agriculture leading to loss of weed wealth. In: Economic policy in India – Time of India. 24th March. [https://timesofindia.indiatimes.com/blogs/economic-policy/herbicide-usage-in-agriculture-leading-to-loss-of-weed-wealth-as-most-weeds-are-medicinal/]

Chopra, R.N., Nayar, S.L. and Chopara, I.C. (1999). Book chapter: Health benefits of *Cynodon dactylon.* In: Council of Scientific and Industrial Research (CSIR), New Delhi. pp. 88-89.

Devi, K.Y., Devi, M.H. and Singh, P.K. (2017). Survey of medicinal plants in Bishnupur District, Manipur, North Eastern India. International Journal of Applied Research, 3 (4): 462-471.

Duarte, M.C.T., Figueira, G.M., Sartoratto, A., Rehder, V.L.G. and Delarmelina, C. (2005). Anti-Candida activity of Brazilian medicinal plants. Journal of Ethnopharmacology, 97 (2): 305-311.

ENVIS. (2022). Mimosa pudica: ENVIS Hub, Kerala. Kerala Environmental Information system. [http://www.kerenvis.nic.in/Database/Mimosapudica_1292.aspx]

Gharde, Y., Singh, P.K., Dubey, R.P. and Gupta, P.K. (2018). Assessment of yield and economic losses in agriculture due to weeds in India. Crop Protection, 107: 12-18.

Gubbiveeranna, V. and Nagaraju, S. (2016). Ethnomedicinal, phytochemical constituents and pharmacological activities of *Tridax procumbens*: A review. International Journal of Pharmacy and Pharmaceutical Sciences, 8 (2): 1-7.

Gunaseelan, V.N. (1987). Parthenium as an additive with cattle manure in biogas production. Biological wastes, 21 (3): 195-202.

Henderson, R.J.F. (1974). *Solanum nigrum* L. (Solanaceae) and related species in Australia. Contributions from the Queensland Herbarium Number, 16: 1-78.

Ji, G., Shukla, S.K., Dwivedi, P., Sundaram, S., Ebenso, E.E. and Prakash, R. (2012). *Parthenium hysterophorus* plant extract as an efficient green corrosion inhibitor for mild steel in acidic environment. International Journal of Electrochemical Science, 7 (10): 9933-9945.

Kiritikar, K.R. and Basu, B.D. (1987). Indian Medicinal Plants. Vol. 3. (2nd Ed). pp. 2061-2062. International book distributors, Dehradun, India.

Kumar, D.A. (2012). Rapid and green synthesis of silver nanoparticles using the leaf extracts of *Parthenium hysterophorus*: A novel biological approach. International Research Journal of Pharmacy, 3 (2): 169-172.

Kumar, S., Masto, R.E., Ram, L.C., Sarkar, P., George, J. and Selvi, V.A. (2013). Biochar preparation from *Parthenium hysterophorus* and its potential use in soil application. Ecological Engineering, 55: 67-72.

Kumari, K. and Saggoo, M.I. (2015). Traditional and Ethnomedicinal uses of some grasses (Poaceae) of Kinnaur and Himachal Pradesh, India. Annals of Plant Sciences, 4 (10): 1195-1198.

Lansdown, R.V. (2013). *Echinochloa colona*: The IUCN Red List of Threatened Species. [https://www.iucnredlist.org/fr]

Martin Quintal, Z., Moo Puc, R., Gonzalez Salazar, F., Chan Bacab, M.J., Torres Tapia, L.W. and Peraza Sanchez, S.R. (2009). In vitro activity of *Tridax procumbens* against promastigotes of Leishmania mexicana. Journal of Ethnopharmacology, 122 (3): 463-467.

Meenakshi, C. (2019). Bhringraj (*Eclipta alba*) properties, benefits, uses and dosage. Planet Ayurveda. [https://www.planetayurveda.com/library/bhringraj-false-daisy-eclipta-alba/]

Melillo, G. (1987). Capsules containing the active principle of an allergen and process for their preparation. United States Patent: 4681752. July, 21, Washington.

Nagori, B.P. and Solanki, R. (2011). *Cynodon dactylon* (L.): A Valuable Medicinal Plant. Research Journal of Medicinal Plants, 5 (5): 508-514.

Quattrocchi, U. (2006). CRC World dictionary of grasses (1st Ed.): common names, scientific names, eponyms, synonyms and etymology. pp. 1-2408. CRC Press, Taylor and Francis. Boca Raton, USA.

Rajiv, P., Rajeshwari, S. and Venckatesh, R. (2013). Bio Fabrication of zinc oxide nanoparticles using leaf extract of *Parthenium hysterophorus* L. and its size dependent antifungal activity against plant fungal pathogens. Spectrochimica Acta Part A: Molecular and Biomolecular Spectroscopy, 112: 384-387.

Ramachandran, J. (2008). Herbs of siddha medicine/the first 3D book on herbs. pp. 1-156. Murugan Pathipagam, Chennai, India.

Ramachandra Prasad, T.V. (2011). Proceedings of the Annual Group Meeting: All India Coordinated Research Project on Weed Control on 28.02.2011 to 01.03.2011. pp. 1-18.

Rastogi, R.P. and Mehrotra, B.N. (1991). Compendium of Indian medicinal plants. Central Drug Research Institute, Lucknow and publications and Information Directorate, New Delhi.

Saijo, R., Murakami, K., Nohara, T., Tomimatsu, T., Sato, A. and Matsuoka, K. (1982). Studies on the constituents of *Solanum* plants II. On the constituents of the immature berries of *Solanum nigrum* L. Journal of the Pharmaceutical Society of Japan, 102 (3): 300-305.

Saikia, A.P., Ryakala Venkat, K., Sharma, P., Goswami, P. and Bora, U. (2006). Ethno botany of medicinal plants used by Assamese people for various skin ailments and cosmetics. Journal of Ethnopharmacology, 106 (2): 149-157.

Saini, S. and Saini, K.K. (2020). *Chenopodium album* Linn: An outlook on weed cum nutritional vegetable along with medicinal properties. Emergent Life Sciences Research, 6 (1): 28-33.

Selvanayagam, Z.E., Gnanavendhan, S.G., Balakrishna, K. and Rao, R.B. (1995). Anti-snake venom botanicals from ethnomedicine. Journal of Herbs, Spices and Medicinal Plants, 2 (4): 45-100.

Shaily, S. and Sarmad, M. (2018). Pharmacological properties of *Cardiospermum halicacabum*: A review. Journal of Environment, Science and Technology, 4 (2): 48-52.

Siebert, T.E., Wood, C., Elsey, G.M. and Pollnitz, A.P. (2008). Determination of rotundone, the pepper aroma impact compound in grapes and wine. Journal of agricultural and food chemistry, 56 (10): 3745-3748.

Simpson, D.A. and Inglis, C.A. (2001). Cyperaceae of economic, ethnobotanical and horticultural importance: A checklist. JSTOR digital library-Kew Bulletin, 56 (2): 257-360.

Soltan, M.M. and Zaki, A.K. (2009). Antiviral screening of forty-two Egyptian medicinal plants. Journal of Ethnopharmacology, 126 (1): 102-107.

Sumitra, S. and Parul, S. (2018). Medicinal potential of weed *Echinochloa colona* (L.) Link: A review. International Journal of Pharmacognosy and Phytochemical Research, 10 (3): 98-102.

Traci Watson. (2014). Ancient people achieved remarkably clean teeth with noxious weed. In: National Geographic. July, 16.

Yamashita, O.S., Azevedo, G.W.P., Peres, W.M., David, G.Q., Carvalho, M.A.C. and Koga, P.S. (2017). Seedling production of fruits and ornamental species to the use of weed hormone (*Cyperus rotundus*). Nucleus, 14 (1): 279-288.

Yu, H.H., Lee, D.H., Seo, S.J. and You, Y.O. (2007). Anticariogenic properties of the extract of *Cyperus rotundus*. The American journal of Chinese medicine, 35 (3): 497-505.

14

Potential of Vermicompost Organic Manure in Integrated Farming System

***Nisha Arya*[1], *Sudhanand Prasad Lal*[2] and *Naman Kumar Mishra*[3]**

[1]*Department of Family and Community Resource Management, Faculty of Family and Community Science, The Maharaja Sayajirao University of Baroda, Vadodara, Gujarat*

[2]*PG Department of Extension Education, Dr. Rajendra Prasad Central Agricultural University, Pusa, Samastipur, Bihar*

[3]*Senior Executive - Marketing Mankind Agritech Ltd. 262, Phase-3 Okhla, New Delhi*

Abstract

'Earthworms are the best friends of farmers'; no doubt vermicomposting clearly illustrates this concept. Vermicomposting is easy as well as economical but it also offers a lot of environmental advantages. The process is environmentally friendly and involves a completely natural process. It also addresses one of the most alarming problems, which is waste management and minimisation. The present study scrutinizes the economic benefits, entrepreneurship opportunities, and challenges related to the increasing vermicompost-based entrepreneurship prospects. Vermicompost production process involves activities for locally focused, resource-efficient food production that are incorporated into the production of organic fertiliser. In the agriculture sector, with minimum investment the production of vermicompost organic fertilizer is a major component of agri-entrepreneurship. In the vermicompost production industry, entrepreneurs may encounter challenges such as a lack of awareness regarding techniques used in vermicompost production, a lack of marketing experience, inadequate resources for training, financial difficulties, dealing with earthworm survival, etc. To overcome these challenges various government and non-government organisations are developing awareness and training programmes and providing financial schemes to help agripreneurs start their business in the field of vermicompost organic manure.

***Keywords*:** Agri-entrepreneurship, Challenges, Economics, Opportunities, Vermicompost.

Introduction

Integrated farming systems (IFS) generate a large amount of organic waste matters. In general, these residues should can be reprocessed appropriately within agri-business systems to fulfil the nutritional and energy demands in the enterprises (IFS) in order to increase productivity. To reduce the heavy consignment of organic waste matter from the Integrated farming systems, The Multi Resource Vermicompost (MRV) was defined by the IFS as an organic fertiliser made from waste produced in crop production, farm animals, agriculture, and vegetation through bio-oxidative degradation processes facilitated by earthworms and microorganisms in the soil.

Vermicompost is a catch phrase for the agri-entrepreneurs because its quality to stop two gaps with one bush. Vermicompost technology deals with the organic waste that's convert into an organic waste into a nutrient rich compost naturally. Vermicompost is a composting product or method that uses several worm species to produce a composite of decomposed vegetable or organic waste, bedding materials, and vermicast. Earthworms eat this decomposed waste and excrete it indigesting form. Earthworms consume organic waste and excrete "vermicast," which are nitrate- and mineral-rich deposits of phosphorus, magnesium, calcium, and potassium. These improve soil quality and are used as fertilisers, this is the reason Prosper (2020) referred the compost as "Black Gold".

Vermiculture refers to worm farming used to produce vermicompost. Proper management, minimisation, and recycling of these wastes in agriculture should be a part of our way of life in order to protect the environment for future generations, strengthen and sustain ecological integrity and community, and focus on promoting the revolution of discoveries concerning the golden fertiliser in farming *via* worms. The finest solution for healthy agriculture is vermicompost (El-Jawaher *et al*., 2019). In addition to higher levels of organic carbons and readily available plant nutrients, organic manure made from a variety of farm resources also contains a number of plant growth-promoting compounds (such as enzymes, vitamins, and hormones) and advantageous micro-organisms that promote healthy and better plant growth. Likewise, it has a number of inert qualities that can lessen soil-borne infections and plant illnesses.

Importance of Vermicomposting in Integrated Farming System

Earthworms are known as "farmers buddies" because they increase the extent of clay and nitrogen molecules in topsoil, hence improving soil fertility in the Integrated Farming System (IFS). The agriculture sector is thriving on the use of earthworms to replenish depleted soil in fields and home gardens.

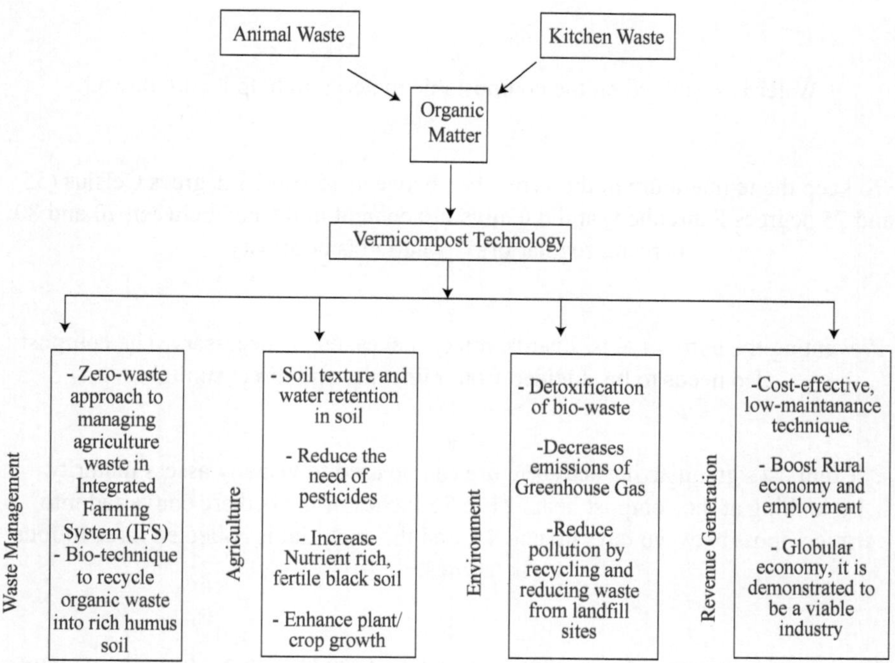

Preparation of Vermicompost

The epigeic species of earthworms, i.e., *Eisenia fetida* (red wigglers), *Eudrilus eugeniae* (African nightcrawler), and *Perionyx excavatus* (Indian blues or composting worms) are promising worms used for the production of vermicompost.

The following steps are needed for preparation of vermicompost –

Collection of waste from the households or farms and keep them into the clean and shady place.

↓

To speed up the decomposition process, chop the waste into small chunks using a waste thresher machine and leave it for 7 to 10 days.

↓

A heap of between 0.6 and 0.9 metres in height can be created by layering of decomposed waste and cow dung.

↓

Until the bed reached the desired height of 30 inches (2'), chopped decomposed waste and slurry cow dung were added alternately and it becomes the heap of 120"×60"×30" (8'×3'×2') and kept it for 4-5 days.

↓

Water is sprinkled on the compostable material to help it cool down.

↓

To keep the temperature in the vermi-bed between 15 and 25 degrees Celsius (55 and 75 degrees Fahrenheit) and the moisture content in the bed between 70 and 80 percent, regular drip irrigation is necessary.

↓

Preventing the entry of ants, lizards, mice, snakes, etc. is necessary. The compost also needs to be shielded from rainwater and direct sunlight.

↓

Compost's granular surface structure can be used to visually assess maturity by looking at the compost heap. 21 to 23 inches of the bed are converted into vermicompost between days 40 and 45, and this material is collected three to four times.

↓

Earthworms, cocoons, and foreign material are used to separate from the manure using one of two methods: manually or mechanically sieving the compost.

↓

The separated manure is ready to be packaged, sold, and applied to farmland as fertiliser.

Nutrient Elements of Vermicompost *vs.* Compost

In terms of the chemical composition of soil, the most important chemical property is its organic matter content, which regulates the major properties of soil. Though research on soil health degradation is vast, its direct relationship with declining food crop output remains a contentious issue. However, due to seamless farm operations, no root residue enters the soil system in the field. In Mother Nature's soil laboratory, a variety of organisms (both macro- and micro-) have the potential to transform organic waste into useful resources that contain plant nutrients and minerals.

According to Srinivaso *et al.* (2011), the following chart shows the effective increment of nutrients (%) in inorganic manure by including worms for making vermicompost compared to the farmyard manure per kg.

	Nutrients	Compost	Vermicompost (g/kg)
1	Organic Carbon (C)	12%	14%
2	Nitrogen (N)	0.8%	0.51-1.6%
3	Phosphorous (P)	0.4%	0.19-1.02%
4	Potassium (K)	0.5%	0.15-0.73%
5	Calcium (Ca)	2.2%	7.61%
6	Magnesium (Mg)	0.5%	0.568%
7	Sodium (Na)	0.01%	0.06%
8	Zinc (Zn)	0.0012%	0.011%
9	Copper (Cu)	0.0017%	0.0048%
10	Iron (Fe)	1.169%	1.33%
11	Manganese (Mn)	0.041%	0.21%

Potential of Agri-entrepreneurs in Integrated Farming System (IFS): A Meta-Analysis

Due to the marvellous benefits of vermicompost organic fertilizer, farmers have a high demand in the marketplace as it improves the quality of agricultural products is less expensive and replaces harmful chemical fertilisers. With a relatively modest initial investment to convert waste into vermicompost organic fertiliser the market potential of vermicompost technology is already a significant part of agribusiness models all over the world. It is habitually employed in home gardens and pot culture as well. In addition, vermicompost is purchased wholesale by numerous government and non-government organisations dealing with agriculture, forestry, and horticulture. Its demand has increased over the years. Gardening enthusiasts are also looking for vermicompost to promote indoor and outdoor flowering and foliage plants.

According to Thirunavukkarasu *et al* (2022) the analysis of cost-benefit and economic profit showed that the net profit ratio for the first and second harvests was 139.5% and 134.6%, respectively. Using a 15% rate of interest, the net present worth (NPW) of the cost and benefit was estimated to be ₹7.63 lakh and ₹11.94 lakh, respectively. This resulted in a benefit-cost ratio (BCR) of 1.56, which allows for the possibility of earning a respectable sum. The Nainer (2019), in the newspaper “The Hindu” published about Ms. Dheeran, who began her entrepreneurial journey in vermicompost in August 2019 with an investment of 1 lakh and holds a Masters in Chemistry and a B.Ed. from the Government College for Women in Kumbakonam. Under the trade name “Soil Spirit,” she hired 4 people and began producing roughly 1 tonne of vermicompost-based planting mix per week that is strengthened with soil and cow manure. For her product, she just registered the trademark “Arboreal”.

According to the report by Singh (2022), Black Gold-Vermicast used as a biofertilizer fascinated the 51-year-old farmer Parvin Srivastava from Varanasi, UP. He developed a network of 400 farmers and adopted vermicompost technology to make organic fertiliser generating not only revenue for themselves and creating employment opportunities for others without leaving their homes. Pandit (2021) reported on Sana Khan, the owner of SJ Organics from Meerut UP, who is enthusiastic about vermicomposting. Engineer Sana was spellbound by earthworms when she worked on vermicomposting as her college project leading her to set up her own agri-entrepreneurship. In 2015, Sana began to earn money and expand her business. By 2020, she was receiving 500 tonnes of waste each month, generating 150 tonnes of vermicompost and earning revenue of ₹1 crore in a year. Sana had started to generate revenue and grow her company in 2015.According to Chakrabarty (2019), comparing vermicompost to traditional compost, the fraction of macro and micronutrients is two times higher. Depending on the type of organic waste, earthworm inhabitants, and environmental variables, the product is ready to use as fertilizer between 45-60 days. With this concept, in a year there are 5 to 6 probable vermicompost cycles. Moreover, it has high market potential across an inclusive choice in producing crops, including agricultural, horticultural, ornamental, and vegetable. They also mentioned that vermicompost business has high-end market results in both village and metro regions because of the desirability of naturally produced farm products. A successful marketing strategy can involve direct sales and marketing of vermicompost by producers to nearby customers *such as* local farmers, gardeners, plant nurseries and home makers. There are also cases of marketing through dealers and cooperatives. Direct sales of vermicompost by producers to neighbourhood customers are seen as a successful marketing strategy. However, there are also a few cases of marketing through dealers and cooperatives. In comparison to marketing through dealers, where the producers' share was lower at 71%, marketing vermicompost through cooperatives had a greater producers' share of 88%.

According to the Verified Market Research (2022) report, it was estimated that the global vermicompost market expanded at a CAGR of 16.74% from 2020 to 2027, increasing from a value of USD 63.55 million in 2019 to USD 222.42 million by 2027. This was possible due to the intensifying demand for vermicompost organic fertiliser for various uses *such as* agriculture, horticulture, landscaping, nurseries, plant or botany laboratories and home gardening. This showed the worldwide vermicompost market has experienced substantial expansion in recent years. Vermicomposting technology is progressively in demand from private laboratories, adopted by universities, various organisations and also schools for use in research projects and instructional purposes, which is fuelling the market's potential of vermicompost organic products in India also.

Many Krishi Vigyan Kendras i.e. Rudrapur, Nizamabad (Shwetha *et al*., 2020) by Sivasagar District of Assam (Bardoloi *et al*., 2020) University also adopted vermicomposting technology to kill two birds with one stone.

SUKHET Model: Use of Vermicomposting Technology

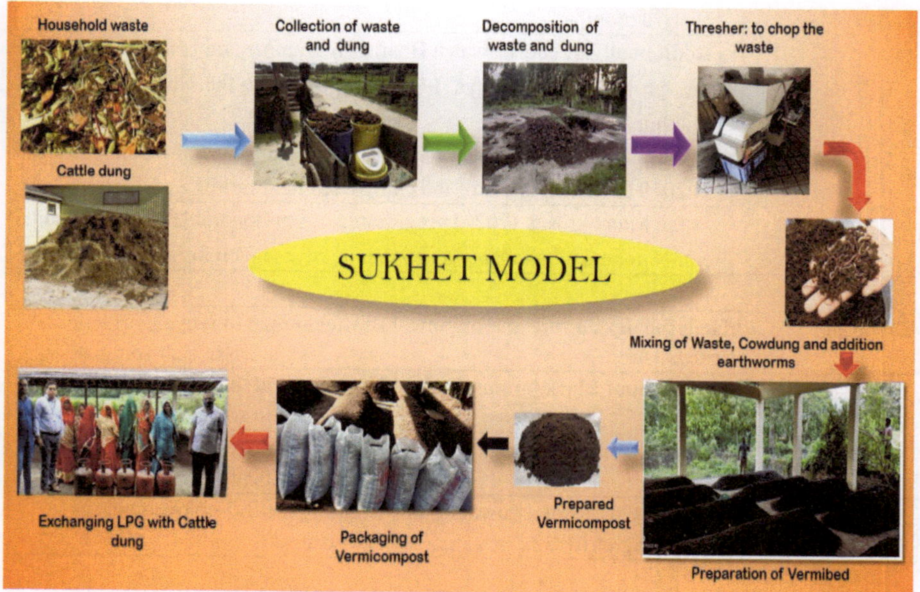

Four-fold benefits of SUKHET Model

Sukhet Model	Sukhet Model is a project initiated by the agricuture scientists of Dr Rajendra Prasad Agriculture University, Pusa, Samastipur Bihar to check pollution.
	The experiment was carried out in a village called Sukhet in the district of Madhubani, Bihar, hence the name Sukhet Model.
	This project is designed an intriguing way to recycle the growing amount of organic waste and cut back on fertiliser use.
	Additionally, using composts on a big scale is an excellent approach to boost the soil's organic matter content, which is essential for their long-term fertility.

	Also, the Sukhet Model helps to control the pollution by getting rid off the agriculture and animal waste from landfiling and convert the waste into rich-humus black fertile soil.
	The Sukhet Model design includes the collection of garbage from door to door and dung and then recycle them to make the vermicompost (organic manure).
	Waste is act as the raw material for the model. Hence, in exchange of the waste and cattle dung every household is provided LPG cylinder in every 2 months.
	Additionally to boost Swatch Bharat programme and benefiting farmers, this project also employs 14 to 15 people in the Sukhet village of Madhubani, Bihar.
	The agriculture department of Pusa, Bihar introduced the Sukhet Model in 2019 as a component of the climate resilient farming programme.
	In the Mann ki baat (2021) programme of our honorable Pradhan Mantri Mr. Narendra Modi also praised the intiatives taken in Sukhet Model.

Disposal	Managing agriculture and organic waste waste	Bio-technique to recycle organic waste into nutrient rich compost
Improve	Conversion of Black humus fertile soil	Enriched with beneficial micro-organisms in soil
Pollution	Emit smoke and carbn dioxide	Helps to improves the quality of air and land to beat the pollution
Sustainability	Monetizing waste and cowdung to pay refilling of LPG	Creating local Employement opportunities

Constraints for Adopting Vermicompost Technology

Worms are generally not fussy eaters, according to Prosper (2020). They may eat fruit and vegetable scraps, cereals, lentils, nuts and breads, egg shells, coffee beans, lawn or grass clippings, and green/dead leaves. Meat, dairy, and oils should not be fed to worms since they will produce a powerful odour and may attract bugs. It is also not advisable to feed worms citrus peels, onions, or garlic. The lack of training facilities, an unorganised vermicompost market, a lack of extension contacts, a lack of government support, a lack of scientific knowledge, a lack of pertinent material, insect and disease attacks, and a lack of extension contacts are the top barriers (Kalita *et al.*, 2019). Lowering the price of chemical fertilisers, which are now less expensive when compared to the price of vermicompost for providing an equivalent amount of nutrients. However, these drawbacks are somewhat similar to those found by Loitongbam *et al.* (2019), who noticed that vermicompost entrepreneurs faced constraints such as ignorance in continuing to improve vermicompost production technology, lack of market information, shortage of training resources, financial difficulties, and challenges in obtaining earthworms. These are the constraints for adopting vermicompost technology among farmers (Sannigrahi, 2016) –

Technical Constraints	• Lack of seriousness. • Lack of formal training. • Lack of skill about improved training. • Labour intensive.
Economic Constraints	• Uninterested youth. • Low cost of chemical fertilizer. • Cost effective but time consuming activity. • Lack of market information.
Psychological Constraints	• Waste handling is a low dignified work. • Many young people are timid or dislike working with earthworms. • Vermicomposting is a more involved process than burning wastes for waste disposal, and it is more difficult to make money from vermicomposting.
Promotional Constraints	• Lack of interconnection/collaboration among different groups of India. • Lack of suitable earthworms for purchase in the neighbourhood. • Rising labour costs hinder farmer's participation. • Lack of laboratory facility for chemical analysis of processed vermicompost and a Quality Certified Agency in Rural India.

Conclusion

The world is stepping up to have sustainable approach especially in the agriculture sector that must fit in three objectives of sustainable growth – environmental, social justice and economic profitability. Vermicompost technology stands on all three goals of a sustainable approach. In order to support the massive eco-friendly technology and earn profits in agriculture, there are several grants, subsidies and credits offered to those who wants to take initiatives such as smallholders, large holders, cooperatives, government and NGOs and farmers' groups to establish vermicompost units in India. The National Mission for Sustainable Agriculture (NMSA), the MGNREGA, the Paramparagat Krishi Vikas Yojana (PKVY), and the Mission for Integrated Development of Horticulture (MIDH) are some of the initiatives that the Government of India uses to assist small farmers or producers in setting up vermicomposting units. Farmers may also form self-help groups or obtain financial assistance from NGOs for small-scale projects. Additionally, several federal banks offer loans to farmers and agriculturalists to help them with their operating capital or investment needs.

References

Bardoloi, A., Bora, R. and Handique, P. (2020). Adoption of Vermicompost Technology for Livelihood Improvement of Farmers in Sivasagar District of Assam. International Journal of Current Microbiology and Applied Sciences, 9(2): 104-108.

Chakrabarty, R. (2019). How to Start Vermicompost Fertilizer Production Business. Retrieved on Sept 16, 2022 from https://www.99businessideas.com/vermicompost-fertilizer-production/

El-Jawaher, A., Bin, D. (2019). Black Gold: The Best alternative in Waste Management via Agriculture. Agri. Res. & Tech: Open Access J. 21(1): 556153.

Jagdish. (2021). Vermicompost business plan in India. Retrieved on Sept 17, 2022 from https://www.agrifarming.in/vermicompost-business-plan-in-india

Jain, S.K. (2021). Mann ki Baat: PM Modi Hails Sukhet Model of Madhubani. Times of India. Retrieved on Nov 16, 2022 https://timesofindia.indiatimes.com/city/patna/pm-modi-hails-sukhet-model-of-madhubani/articleshow/85747314.cms

Kalita, R.R., Das, M.D. and Borah, S. (2019). Constraints faced by farmers in adoption of vermiculture technology. International Journal of Chemical Studies. 7(6): 1942-1944.

Loitongbam, J., Hazarika. S., Choudhury, B.V., Thangavel, B. and Arumugam, B. (2019). Vermicomposting: An entrepreneurship option for the unemployed rural youth of north-eastern India. Conference: National Agri- Business Entrepreneurship Conclave (NABEC) At: ICAR RC NEH Umiam, Meghalaya.

Nainer, N. (2019). Ariyalur homemaker to turn a neat profit with vermicompost venture. The Hindu. Retrieved on Sept 16, 2022 from https://www.thehindu.com/news/cities/Tiruchirapalli/ariyalur-homemaker-to-turn-a-neat-profit-with-vermicompost-venture/article38304037.ece

Nayak, P.K., Nayak, A.K., Shahid, M., Tripathi, R., Panda, B.B., Kumar, A., Kumar, U., Mohanty, S., and Das, S.K. (2021). Mr. Vermicompost: A product of integrated farming system, NRRI Technology Bulletin – 153, NRRI, Cuttack.

Pandita, U. (2021). Fascinated By Earthworms, UP Girl Starts Vermicomposting, Earns ₹1.00 Crore Revenue. Retrieved on Sept 15, 2022 from https://www.thebetterindia.com/249941/how-to-vermicompost-earn-revenue-sana-khan-sj-organics-earthworms-uttar-pradesh-sustainability-c24/

Prosper, J. (2020). Vermicomposting: Making Black Gold. Retrieved on Sept 17, 2022 from https://www.adirondackalmanack.com/2020/03/vermicomposting-making-black-gold.html

Sannigrahi, A.K. (2016). Major Constraints in Popularising Vermicompost Technology in Eastern India. Modern Environmental Science and Engineering, 2(2): 123-133.

Singh, B. (2022). Varanasi Fulfilling natural farming vision of PM Narendra Modi with Black Gold. The Times of India. Updated: Aug 17, 2022, 10:25 IST. Retrieved on September 17, 2022 from https://timesofindia.indiatimes.com/city/varanasi/fulfilling-natural-farming-vision-of-pm-with-black-gold/articleshow/93605964.cms

Sukhet Model https://edukemy.com/current-affairs/gazette/2021-09-01/sukhet-model#:~:text=Sukhet%20Model%20is%20named%20after,into%20vermicompost%20(organic%20manure).

Support from Government: Schemes providing assistance. Retrieved on Sept 14, 2022 from https://vikaspedia.in/agriculture/agri-inputs/bio-inputs/support-from-government

Swetha, M., Balazzii Naaiik, R.V.T., Vijay Kumar, P., Rajkumar, B., Vijayalakshmi, D. and Bhavyamanjari, M. (2020). Adoption of vermicompost technology. Indian Journal of Entomology. 82(1): 32-35.

Thirunavukkarasu, A., Nithya, R., Makesh Kumar, S., Priyadharshini, V., Prem Kumar, B., Premnath, P. Sivashankar, R. and Bose Sathya, A. (2022). A business canvas model on vermicomposting process: key insights onto technological and economic aspects. Bioresource Technology Reports. 18: 101119.

Verified Market Research (2022). Vermicompost market size and Forecast. Retrieved on Sept 15, 2022 from https://www.verifiedmarketresearch.com/product/vermicompost-market/.

15

Enriching Integrated Farming System Through Precision Farming to Boost Entrepreneurship – Application of IoT and AI

Vicky Yadav and Sudhanand Prasad Lal

PG Department of Extension Education, Dr. Rajendra Prasad Central Agricultural University Pusa, Samastipur, Bihar

Abstract

Precision Farming works on the concept of the right input, in the right amount, at the right time, which means no wastage of input, and IFS also work on the same concept. Once the output is used, input for another enterprise, so if we see both are working on the same idea, it means no wastage. It is a good farming method for Indian farmers because it generates entrepreneurship opportunities for farmers. Nowadays, the current trend in the Indian job market is "be a job creator, not a job seeker." IFS can serve as a good alternative in this regard. If we include some precision farming components in IFS, these challenges can be mitigated, but the installation cost of these components is very high, so the Government has to provide some subsidies for these components. PF components reduce the excessive use of agrochemicals, so climatic conditions are improved. These components are also helpful for farmers' health and the environment because only agriculture contributes to 10-12 % of GHG emissions. IFS improve farmers' family conditions because it gives them a chance to earn money from different sources.

Keywords: Entrepreneurship, GPS, GIS, IFS, Natural Farming, Precision farming, Sustainable agriculture.

Introduction

Precision farming is a site-specific farming method that works on a large scale, while IFS divides the whole land into fragments. Both Precision farming and IFS work on the concept of "*no wastage*". When we install some components of precision farming in IFS, such as remote sensors, yield mapping, SSNM, yield monitors, and some others, the yield and quality of output increase. Initially, farmers may not get much profit from them because their installation cost is very high, but after a while, they earn more profit from this. IFS are good for Indian farmers because many enterprises are connected to each other in IFS, and farmers can get income from different sources. It generates employment opportunities for other farmers. It's good for those with less land, and when we include PF in IFS, entrepreneurship opportunities are generated for educators like farming device installation, operation, maintenance and many others.

Precision Farming

"*Precision farming*" refers to the objective of managing agriculture more effectively. It is a farming method where inputs are used in the required amount, at the needed time, in the proper manner in the field and works on the principle of no wastage. It is also called as site-specific farming method. It is an integration of various technologies like GPS, GIS VRT etc. Precision farming depends on the measurement and understanding of variability. Growers can increase the effectiveness of inputs by better directing them to the precise location where they are required. As a result, the wastage of information is minimised. Precision Farming also helps in integrated crop management (ICM), says Peter Pitkin. The principle of precision farming is the right input, in the right amount, at the right time, in the right manner, and in the right place.

Fig. 1: Agricultural Drone

Fig, 2: Laser Land Leveller

Tibre

TIBRE (Targeted Input for a Better Rural Environment) is a project started for the sustainable use of agricultural inputs. It possesses the question of how we may employ new technology to minimise the negative effects of intensive agriculture on the environment while maintaining or increasing current productivity levels.

Disadvantages of Precision Farming

- The initial installation charges are very high.
- It may take several years to collect sufficient data before installation.
- Farmers may not have sufficient knowledge of the operation and its components.
- Precision agriculture techniques are in the early stages of development, so operating them requires professional knowledge and guidance.

Pathway of Precision Farming

Global Navigation Satellite System (GNSS) is used to provide positioning and timing data to GNSS receivers. The receivers then use this data to determine location. The acquired data such as field history, soil survey, topography, soil sampling, remote sensing and available resources is then stored in storage devices. After that, the processing of this data, imaging and prescription maps are generated. Prescription maps are generated from a range of geo-referenced data, such as soil nutrient levels and historical yields, and provide input rates for defined zones of a field. All these components are used.

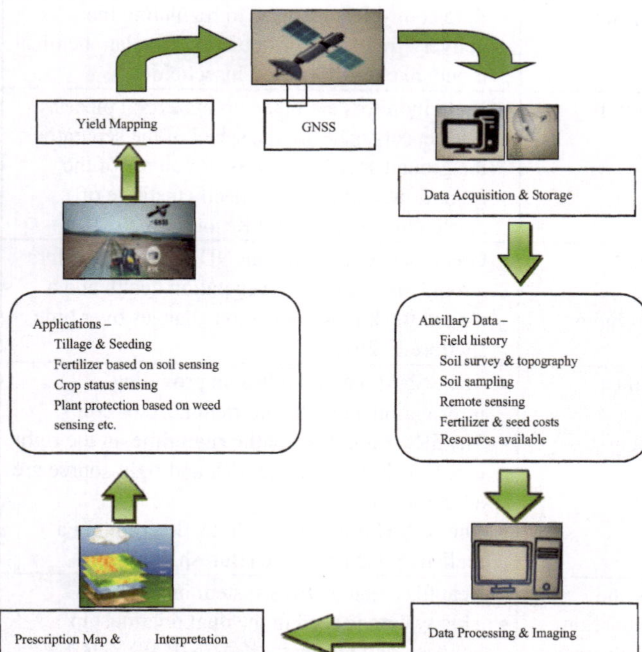

Fig. 3: Pathway of Precision Farming

Table 1: Components of Precision Farming

Abbreviation	Full Form	Uses
GPS	Global Positioning System	• It's used for obtaining data. • Farmers can use GPS to find specific field data locations like soil type, field topography, insect attack, weed infestation, field boundaries, etc. • It offers automatic functionality.
GIS	Geographic Information System	• GIS is used to produce maps. • Computerised GIS maps differ from other maps and have multiple informational layers (e.g. yield of crop, nutrient level of soil, pest attack, rainfall, soil survey etc.). • An agricultural GIS database can offer details on crop production, the field's topography, the type of soil, soil testing, and irrigation (Hakkim, 2016).
VRT	Variable Rate technology	• With the help of variable-rate technology (VRT), various inputs are applied like chemical fertiliser, insecticide, weedicide, fungicide, lime, gypsum and other agricultural inputs (NDSU, 2022).
RS	Remote Sensor	• Remote sensors collect information about an object or area without making direct contact (Rattan *et al*., 2015).
RC	Rate Controllers	• Rate controllers are used to regulating the delivery frequency of liquid or granular chemical inputs like fertilisers and insecticides.
YM	Yield Monitor	• Yield monitors are made up of several pieces. • The sensors measure the speed of the separator, the ground speed, the mass or volume of the grain flow, and the grain itself (Institute of Agriculture and Natural Resources, 2022).
NDVI	Normalised Difference Vegetation Index	• Green Seeker provides an NDVI score. It offers a rough assessment of vegetation health and a way to track how vegetation changes over time (Pettorelli, 2013).
SSNM	Site-Specific Nutrient Management	• The SSNM method relies on providing information about the nutrient need of crops. • The 5Rs: right dose, at the right time, in the right place, with the right method, and right source are addressed by SSNM. • The SSNM processes include: determining a realistic yield objective (Buresh, 2007).
DRIS	Diagnosis and Recommendation Integrated System	• Beaufils created DRIS system in 1973. • This system is used in the dual relationship between two nutrients (K/N, N/P, P/N, etc.) (Beaufils, 1973).

Integrated Farming System

Integrated Farming System is a farming system used to manage an entire farm and provide more sustainable agriculture. The main concept of IFS is that there is no waste and that trash is merely a resource that has been mishandled but could have been used to make something useful. It is a farming system in which many enterprises are interconnected with each other, where the output of one enterprise is used as an input for another enterprise. It reduces input costs, improves soil health, increases output, increases profitability, and generates more income.

Sustainable Agriculture: Sustainable agriculture aims to meet the present generation's requirements without jeopardising the future's resources.

Primary Goals of Integrated Farming System

1) Maximisation of yield,
2) Rejuvenation of the system,
3) Avoid disease, pest and weed population, and
4) Reducing the use of chemicals.

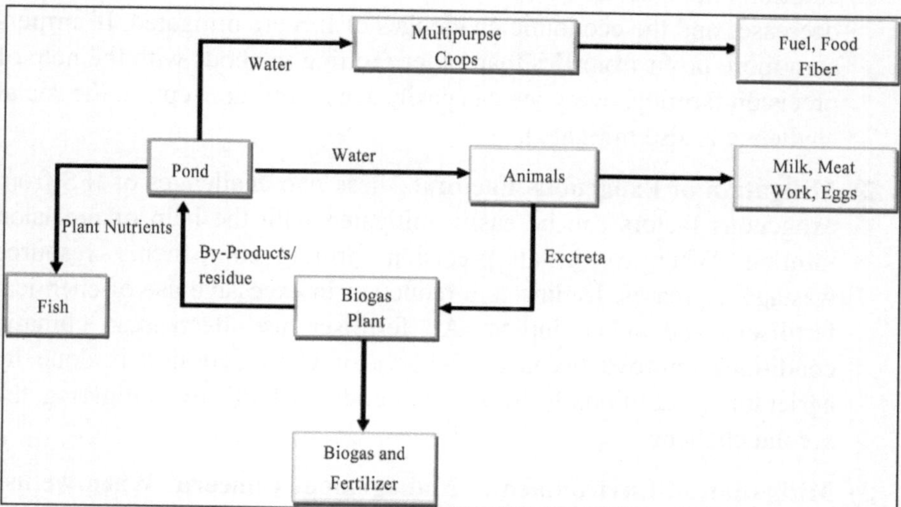

Fig. 4: Different Components and Resources Flow in Integrated Farming System

Challenges of Integrated Farming System

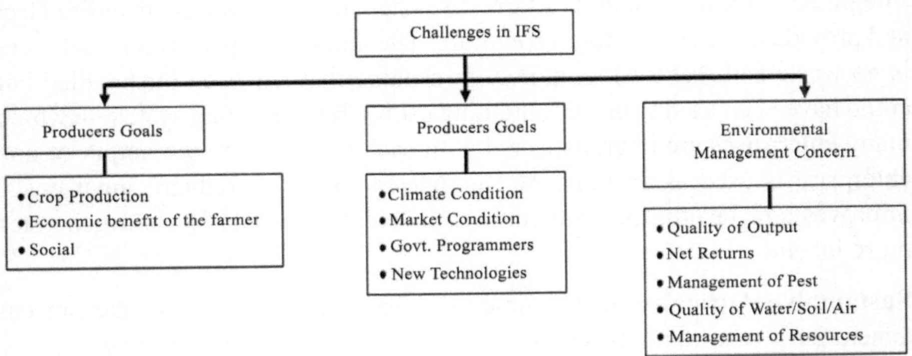

Mitigating the Challenges of IFS through Precision Farming

1) **Mitigation of Producers Goals:** The ultimate goal of a producer is to increase production and profitability. If we install some precision farming components in IFS, then the challenges of IFS can be mitigated. When farming is done with the help of both farming methods, output and profitability increase, and the main concept of PF is followed. In addition, input wastage also decreases. Hence, input cost automatically decreases and the economic challenges of IFS are mitigated. If farmers earn more profit from IFS than other farming methods with the help of precision farming, everyone can easily accept this concept, so the social challenge is also mitigated.

2) **Mitigation of Exogenous Factors:** These two challenges of IFS from exogenous factors can be easily mitigated with the help of precision farming. When we install precision farming components, resource wastage decreases, leading to a reduction in excessive use of chemical fertilisers and other inputs. As fertiliser use decreases, climatic conditions improve because 10-12 % of GHS emission is done by agriculture. Additionally, new technologies of PF are mitigating the second challenge.

3) **Mitigation of Environmental Management Concern:** When we use limited resources in IFS with the help of precision farming components, the quality of output, soil, water, and air quality can be easily improved, and resources can be conserved efficiently. When we use remote sensors, they identify pests easily and manage them. Compared to traditional IFS, IFS with precision farming can increase net returns as input costs decrease, output increases and labour charge decrease. However, the initial installation costs are high.

The Polarity between Integrated Farming System and Precision Farming

Precision farming requires a large area because it operates on a large scale and its initial installation cost is very high. In contrast, IFS divides the entire land into small pieces for different enterprises such as livestock, poultry farms, fish culture and other farming activities. Most Indian farmers are small and marginal, so they cannot afford these precision farming components.

Entrepreneurship Generation through Integrated Farming System

India plays a major role in agriculture worldwide. It has the largest area of wheat, rice, and cotton and is the world's largest producer of milk, pulses, and spices. India also has the world's largest cow herd (buffaloes). About 70% of the rural household of the Indian population, directly or indirectly, depend on agriculture (Uvaneswaran *et al*, 2015; FAO, 2022). At least 80 to 85% of Indian farmers are marginal farmers who do not have sufficient land for survival (Agriculture Census, 2015-16). IFS is a mixed farming system where different enterprises are connected, and the output of one enterprise is used as input for another. In IFS, farmers can get income from various sources. Therefore, IFS is the primary source of income for approximately 65% of the rural population who are reliant on agriculture. Government also focuses on integrated farming system which should act as a key factor in increasing the farmers' income. Therefore, National Mission for Sustainable Agriculture (NMSA) was launched in 2014 by the government specifically for IFS (Department of Agriculture and Farmers' Welfare, 2019). In the current Indian job market, the trend is to "*be a job creator, not a job seeker*". Hence, IFS can serve as a good alternative (The Hindu, 2016).

In an integrated farming system, agriculture, livestock, biogas plant, fish culture, agroforestry and kitchen garden are connected to each other. Managing these activities requires more labour, so employment is generated. If a farmer starts an IFS, he increases profitability and also provides employment to many other farmers. A success story of a Tamil Nadu farmer Shri D. Manivannan started IFS on a 4-acre land and got 75% net returns from IFS (KVK Thiruvannamalai, 2017). IFS also generates employment for women as some activities like kitchen gardening, livestock rearing, and some farming activities are performed by women. If precision farming components are used in IFS, the yield or profitability should increase and entrepreneurship opportunities increase for educators in farming device installation, operation, maintenance etc.

Table 2: Status of Natural Farming

S. No.	State	Area	Percentage
1.	Andhra Pradesh	241,477	35.6
2.	Gujarat	116,666	17.2
3.	Madhya Pradesh	98,484	14.5
4.	Chhattisgarh	84,848	12.5
5.	Kerala	83,333	12.3
6.	Odisha	24,242	3.6
7.	Himanchal Pradesh	12,121	1.8
8.	Rajasthan	7575	1.1
9.	Jharkhand	4545	0.7
10.	Tamil Nadu	3030	0.4
11.	Uttar Pradesh	1515	0.3
	Total		100.0

(Source: Niti Ayog, 2021; PIB, 2021; Dept. of Agriculture, Govt. of Rajasthan, 2021; Agriculture, Farmers Welfare and Co-operation Department, Govt. of Gujarat, 2021; APCNF MIS; Department of Agriculture, Uttar Pradesh).

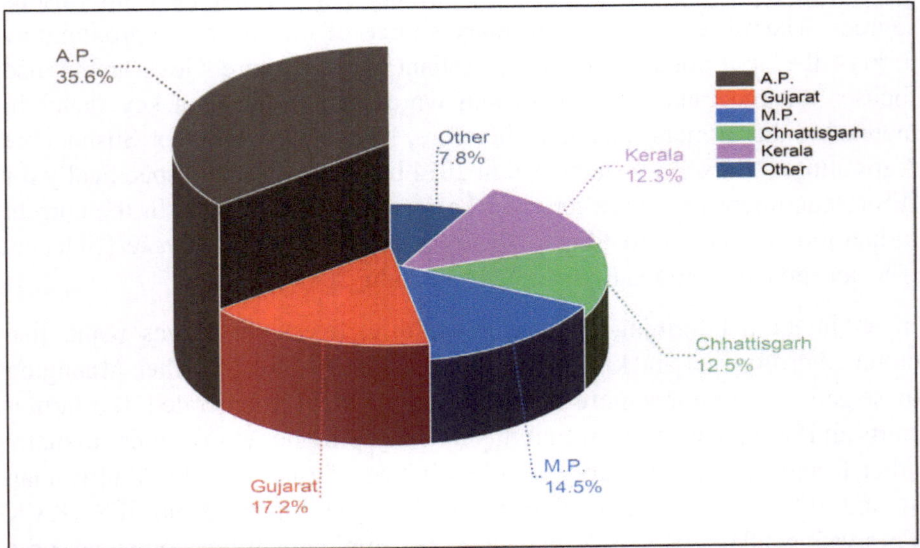

Fig. 5: Top Five States in Coverage of Natural Farming

Conclusion

Precision farming and IFS are based on the same concept of "*no wastage*". When both are present, productivity and profitability increase, generating employment and entrepreneurial opportunities for many farmers. Nowadays, the current trend in the Indian job market is "*be a job creator, not a job seeker*". IFS can serve as a good alternative in this regard. In IFS, farmers get income from different sources. The cost of precision farming components is very high making it unaffordable for farmers to start, so the government needs to provide subsidies on its components. Both contradict each other on some principles; for example, precision farming works on a large scale while IFS divides lands among small enterprises.

References

Agriculture Census, (2015-16). All India Report on Number and Area of Operational Holdings. Agriculture Census Division, Department of Agriculture, Co-Operation & Farmers Welfare, Ministry of Agriculture & Farmers Welfare, Government of India.

Ahmad, L., Mahdi, S. S. (2008). Satellite Farming – An Information and Technology Based Agriculture, Springer Cham, pp. 1-180.

Balkrishna, A., Phour, M., Thapliyall, M. and Arya1, V. (2021). Current Status of Indian Agriculture: Problems, Challenges and Solution. Biological Forum – An International Journal. 13(3): 361-374.

Beaufils, E.R. (1973). Diagnosis and recommendation integrated system (DRIS) (No. Folleto 5743). University of Natal, Department of Soil Science and Agrometeorology, p. 182.

Buresh, R.J. and Witt, C. (2007). Site-specific nutrient management. Fertiliser Best Management Practices – General Principles, Strategy for their Adoption and Voluntary Initiatives vs. Regulations, First edition, IFA, Paris, France, pp. 47-56.

Chaitanya, A. and Adilakshmi, G. (2021). Different techniques and tools in Precision Nutrient Management. Agriculture Letters. 2(7): 2582-6522.

FAO, (2022). India at a glance. Stable URl: https://www.fao.org/india/fao-in-india/india-at-a-glance/en/.

Farmers Weekly. (1994, April 30). Major role for targeted inputs. Stable URL: https://www.fwi.co.uk/news/major-role-for-targeted-inputs.

Hakkim, V.A., Joseph, E.A., Gokul, A.A. and Mufeedha, K. (2016). Precision farming: the future of Indian agriculture. Journal of Applied Biology and Biotechnology. 4(6): 1-7.

Hendrickson, J.R., Hanson, J.D., Tanaka, D.L., Sassenrath, G. (2008). Principles of integrated agricultural systems: Introduction to processes and definition. Renewable Agriculture and Food Systems. 23(4): 265-271.

Innazent, A. and Jacob, D. (2021). Quantitative farming system typologies for targeted agricultural development: A review. The Pharma Innovation Journal. 10(8): 193-198.

Institute of Agriculture and Natural Resources (2022). Yield Monitoring and Mapping. University of Nebraska, Lincoln. Stable URL: https://cropwatch.unl.edu/ssm/mapping.

Kopp, V.A. (1988). Applying the diagnosis and recommendation integrated system (DRIS) to Fraser fir Christmas trees, Master's Thesis, Virginia Polytechnic Institute and State University, Blacksburg, Virginia.

KVK Thiruvannamalai, (2017). A Success story on Integrated Farming System, Annual Report 2016-17, KVK Thiruvannamalai, Tamil Nadu.

Naqvi, S.N.H. (1994). Developing Countries and the Uruguay Round Agreement. Journal of Cooperation among Islamic Countries, 15(1-2), 91-112.

NDSU (2022). Variable Rate Technology, North Dakota State University.

NMSA (2019). Integrated Farming System, National Mission for Sustainable Agriculture (NMSA), Department of Agriculture and Farmers' Welfare, Govt. of India.

Pettorelli, N. (2013). The normalised difference vegetation index, Oxford University Press (1st Edition), New York, USA.

Ram, T., Lohan, S.K., Singh, R. and Singh, P. (2020). Precision Farming: A New Approach, Daya Publishing House, Astral Publication, New Delhi.

Serra, A.P., Marchetti, M.E., Bungenstab, D.J., da Silva, M.A.G., Serra, R.P., Guimarães, F.C.N., do Amaral Conrad, V. and De Morais, H.S. (2013). Diagnosis and Recommendation Integrated System (DRIS) to Assess the Nutritional State of Cotton Crop in Brazil. American Journal of Plant Sciences. 1(1): 129-146.

The Hindu, (2016). Be a job creator, not a job seeker (16th October, 2013). Stable URL: https://www.thehindu.com/news/national/tamil-nadu/%E2%80%9CBe-a-job-creator-not-a-job-seeker%E2%80%9D/article16072788.ece.

Uvaneswaran, S.M. and Keerthana, S. (2015). Integrated farming system (IFS): a new entrant for entrepreneurs. Pezzottaite Journals. 2(4): 2279-0918.

Zhang, N. (2022). Modern Approaches to Smart Farming via Precision Agriculture. Agrotechnology. 11(3): 1000252.

16

Entrepreneurial Skills of Small and Marginal Farmers through Integrated Farming System

A. Devivaraprasad Reddy[1], T. Vijaya Nirmala[1], E. KarunaSree[1&3] CH. Balakrishna[2], K. Venkata Subbaiah[1], G. Shali Raju[1] V. Deepthi[3], M. Raghavendra Reddy[1], K. Venkata Satish[1] and J.V. Prasad[4]

[1]*Krishi Vigyan Kendra, Dr. Y.S.R. Horticultural University, Venkataramannagudem, West Godavari District, Andhra Pradesh*
[2]*Krishi Vigyan Kendra, Acharya NG Ranga Agricultural University Amadalavalsa, Srikakulam District, Andhra Pradesh*
[3]*Administrative Office, Dr. Y.S.R. Horticultural University Venkataramannagudem, West Godavari District, Andhra Pradesh*
[4]*ICAR-Agricultural Technology Application Research Institute (ATARI) Zone-10, Hyderabad, Telangana*

Abstract

India is predominantly an agrarian nation, with agriculture as the primary economic activity and livelihood source for most Indian households. Many small-scale and economically disadvantaged farmers commonly employ monocropping. There exists a correlation between low agricultural yields, diminished income, delayed monsoon, pest and disease vulnerabilities, and various other risk factors. Integrated farming systems (IFS), which involve the integration of multiple crops and enterprises, have emerged as a contemporary solution for augmenting income. Given the escalating demand for animal products and the mounting pressure on land resources, ensuring the judicious utilisation of feed attributes, including grain deposits, is imperative. The Integrated Farming System (IFS) is primarily concerned with optimising farmer profitability through enhancing and integrating agricultural properties, recycling waste materials, and reducing reliance on chemical inputs. One notable aspect of contemporary agricultural practices involves the enhancement of farming standards. Monocropping is associated with substantial risk, resulting in limited crop productivity. The challenges encompass inadequate remuneration and occasional ambiguity surrounding income. Efforts have been undertaken in five areas of the homestead property to support the enterprise, including the cultivation of paddy, fish, maize, and feed grass and the management of dairy and poultry. The empirical evidence supports the notion that the expansion of agribusiness is associated with enhanced farmer efficiency within the designated development zone, while concurrently decreasing human resources. The implementation of diversification in cropping and integrated farming systems has been found to yield significant benefits in improving livelihoods and socioeconomic status.

Keywords: Integrated farming system; Diversification in agriculture; Farming system.

Introduction

Agriculture plays a significant role in the economic development of rural networks in India and constitutes a substantial proportion of the nation's gross domestic product (GDP). Despite experiencing a decline in its GDP share, the agricultural sector in India continues to hold significant importance due to its modernisation efforts, which have led to a substantial boost in government GDP. In the present scenario, there is an urgent requirement for a modification in the agricultural framework within the legislative body to effectively address the challenges encountered by rural areas amidst the evolving domestic and global financial environment. Additionally, measures must be taken to enhance the quality of life for farmers.

The Integrated Farming System (IFS) employs the concept of "integrated farming" to delineate the practice of integrated farming in contrast to monocropping. The domains encompassed within this classification encompass horticulture, agriculture, animal husbandry, fisheries, and related disciplines. Implementing an integrated network of enterprises within the IFS system enables the utilisation of surplus or unproductive output from one segment as a resource for another, resulting in reduced input expenses and enhanced operational efficiency or revenue generation. Consequently, these programmes function in conjunction as a complex network of systems to minimise waste (Anon 2010; Chakrabarti *et al*., 2014; Soni *et al*., 2014; Reddy *et al*., 2018).

The concept of diversification has been extensively employed in agricultural production over the past decade. Rural growth entails the conversion of assets from a combination of low-value items to a combination of high-value products. The primary sectors encompassed in this context are horticulture, dairy, poultry, and aquaculture. Diversification has been proposed as a potential solution for addressing challenges in the agro-industry (Reddy *et al.*, 2019). Diversification has conventionally been regarded as a form of agricultural practice wherein farmers cultivate a variety of crops on their land. Over the previous decade, there has been a growing utilisation of the term "diversification" to denote the broadening of the high-value crops sector. Moreover, cultivating value-added crops is a crucial strategy for enhancing crop productivity and augmenting the economic viability of the agricultural sector. The quantification of crop diversification is based on the economic returns generated by value-added crops. This assessment considers the efficient utilisation of land and human capital, ultimately contributing to increased agricultural productivity. According to Reddy *et al.* (2019), implementing crop diversification has enhanced global crop output in terms of efficiency, quantity, and economic value across diverse agro-climatic conditions. Several potential approaches exist for integrating India's existing rice and wheat planting practices into agricultural systems. Horticulture encompasses various practices such as vegetable planting, floriculture, fruit cultivation, mushroom farming, and the production of medicinal and herbal plants. The utilisation of apiaries also facilitates diverse possibilities, optimising the allocation of human capital, enhancing productivity, and promoting environmental conservation. Moreover, the implementation of fisheries-based systems, such as fish ponds, horticultural and agricultural crops, dairy farming, poultry farming, goat farming, pig farming, and duck farming, has been found to enhance farm revenues and create employment opportunities (Reddy *et al.*, 2018; Reddy *et al.*, 2019). The current chapter focuses on the entrepreneurial skills acquired for enhanced income generation through an integrated farming system approach.

Horticulture

The exponential increase in population is giving rise to novel concerns and complexities for the agricultural sector in India, including issues related to limited availability of land, scarcity of water resources, air pollution, and urbanisation. These challenges impose a burden on the advancement of efficient horticultural systems. In the context of India, it is observed that a significant majority, specifically over 86 percent, of individuals engaged in agricultural activities are categorised as small and marginal farmers. These farmers possess land holdings that are less than 1.2 hectares in size. Due to the continuous growth of the population and the concurrent decline in per capita land availability, the potential for horizontal horticultural expansion is severely limited. Small-scale agricultural operations are often vulnerable to the adverse effects of natural calamities, including intense precipitation, flooding, and prolonged periods of drought. Consequently, these events pose significant risks to farming activities conducted on smaller plots of land. To ensure consistent revenue generation for farmers, it is imperative to employ efficient vertical and horizontal spacing techniques, alongside effective agricultural practices. Implementing an Integrated Farming System (IFS) has been identified as a viable approach to promote sustainable income generation, employment opportunities, livelihood enhancement, and nutritional security among small and marginal farmers (Gururaj *et al*., 2017).

The implementation of an integrated farming system has the potential to enhance productivity by optimising resource utilisation through recycling. The IFS system encompasses a horticulture-based integrated farming system that additionally facilitates income augmentation. The cultivation of fruit, vegetable, plantation, and flower crops collectively contributes to the augmentation of economic prosperity, subjective well-being, consistent financial earnings, and nutritional stability. According to Kashyap *et al*. (2015), various horticultural methods, especially those related to vegetable cultivation, enhance income and productivity within the agricultural system. Furthermore, implementing a geographically targeted horticulture-based agricultural system that emphasises the cultivation of appropriate horticultural crops, including plantation crops, fruit crops, vegetable crops, and flower crops, would result in increased productivity, profitability, and long-term viability.

Table 1: Vegetable based integrated farming system for West Godavari district of Andhra Pradesh

Major component	**Crops grown**	**Yield (q)**	**Cost of cultivation (₹)**	**Gross return (₹)**	**Net return (₹)**	**B:C ratio**
Kharif season						
0.3 ha	Marigold	8.25	18750	41,250	22,500	2.20:1
0.2 ha	Okra	15.2	12,000	15,200	3200	1.26:1
Rabi season						
0.3 ha	Cabbage	3000 heads	10,000	23,000	13,000	2.30:1
0.2 ha	Cauliflower	2000 flowers	6000	18,000	12,000	3.00:1
Summer season						
0.3 ha	Coriander	1000 bundles	3000	9000	6000	3.00:1
0.2 ha	Okra	13.6	13,500	20,400	6900	1.48:1
Total net profit from 0.5 acre of land in a year					63,600	

According to the data presented in Table 1, the cultivation of marigolds and cool-season crops positively impacted the socio-economic conditions of the agricultural community. In the fiscal year 2019-2020, agricultural practitioners reported a substantial increase in their earnings, ranging from ₹50,000 to ₹ 70,000 derived solely from cultivating a mere 0.5 acres of land.

This represents a notable surge compared to previous years, wherein the same land area yielded a comparatively lower income of ₹20,000 to ₹30,000. Furthermore, it has been observed that the practice of cultivating vegetables using the Integrated Farming System (IFS) model has experienced an increase. Bhardwaj and colleagues (2014) and Bhardwaj and Nandal (2015) reported congruent findings.

Paddy

Paddy or rice is the predominant cereal grain in global consumption, with most of the world's populace, particularly in Asia and Latin America, incorporating it into their diets. Rice cultivation is most optimally suited for countries and regions with low labour expenses and ample precipitation, given its labour-intensive nature and substantial water requirements for irrigation. Nevertheless, it can be cultivated in various locations, including challenging terrains like steep hillsides. Rice straw and other agro-industrial by-products are abundantly available in numerous agricultural nations after the conclusion of harvest seasons. The farm wastes have multiple applications, such as being utilised as feed for ruminant animals, facilitating mushroom cultivation, serving as a source of fuel for heating and biogas production, being employed in manufacturing boards or paper, and contributing to producing organic fertilisers.

Fig. 1: Okra Crop in Vegetable-based Integrated Farming System

Fig. 2: Banana Crop in Vegetable-based Integrated Farming System

Paddy straw is a readily available and cost-effective source of feed for ruminant animals, including buffaloes, cattle, goats, and sheep. Livestock farmers commonly gather and arrange rice straw from their rice farms during periods of scarcity or when access to high-quality roughages is restricted. This accumulated rice straw then functions as a supplementary source of nourishment for their animals. Rice straw is a viable, abundant, and cost-effective feed option for livestock such as cattle, buffalo, goats, and sheep in regions where feed availability is limited or where high-quality forage is scarce. Nevertheless, the ability of the animal to utilise and metabolise nutrients derived from its diet into nutritious food products is primarily influenced by the accessibility and calibre of the forage supplied (Aquino *et al*., 2020).

Fish Culture

The aquaculture industry is experiencing rapid growth and generating substantial financial returns. The enhancement of aquaculture production necessitates the strategic optimisation of resources present in ponds and underutilised water bodies. The primary objective of composite fish culture is to optimise the utilisation of available resources within a pond environment, thereby enhancing the overall productivity of the pond (Reddy *et al*., 2019). Carp polyculture is advantageous in enhancing output and productivity. Efficient pond production management favours perennial ponds ranging in size from 0.4 to 1 hectare, with a water depth of 2 to 3 meters. Rainwater collection devices can also be utilised for carp cultivation. Loamy or clay loamy soils are optimal when they have a pH range between 6.5 and 7.5. To address the pH condition of the soil, it is recommended to apply a quantity of quick lime or agricultural lime ranging from 200 kg to 1200 kg per hectare, with application intervals of 2 to 3, depending on the current pH level of the soil. A combination of cow dung and gypsum ($CaSO_4$) is employed to ameliorate sodic soils. In addition to soil management, establishing and maintaining water quality criteria are paramount in the cultivation process, as they play a pivotal role in determining the overall success of the pond. In cultural practices, there is a preference for using a light green or brown hue. Aquatic vegetation poses a significant threat to aquaculture operations and can be effectively controlled through various methods, including mechanical eradication, chemical intervention, and biological measures. Predatory and weed fishes can be effectively managed through the utilisation of various techniques, including the application of mahua oil cake @ 2500-3000 kg per hectare, tea seed cake @ 750-1000 kg per hectare, or bleaching powder @ 300-500 kg per hectare, among other viable approaches. According to Shyam and Kumar (2001), dewatering is commonly recommended as the most effective approach for eliminating waste and predatory fish from small-scale ponds.

The composite culture consists of a combination of six species, namely catla (*Catla catla*), silver carp (*Hypophthalmichthys molitrix*), rohu (*Labeo rohita*), grass carp (*Ctenopharyngodon idella*), mrigal (*Cirrhinus mrigala*), and common carp (*Cyprinus carpio*). According to Reddy *et al.* (2019), the fish species in the fish pond can be categorised into three groups based on their feeding habits and the water level maintained. These groups include the surface feeders, such as catla and silver carp, the column feeders, such as rohu and grass carp, and the bottom feeders, such as mrigal and common carp. The ratio of these groups is 3:4:3, respectively. Weeds can be managed through biological methods, such as employing herbivorous fish species like grass carp (*Ctenopharyngodon idella*), *Puntius javanicus*, *Puntius punchellus*, and Gorami. The assessment of stocking density (SD) is determined based on several criteria, including the selection of species, the productivity of the pond, and the size of the pond. Typically, the ponds are populated with a quantity of fingerlings ranging from 5000 to 7000, each measuring between 10 and 15 centimeters in length. In the context of zero-point culture, it is common practice to introduce advanced fingerlings weighing between 150 to 250 grams, typically in quantities ranging from 500 to 1000 individuals (Ramakrishna *et al.*, 2013). During stocking, it is common practice to subject fish to a brief bath in a solution containing either 3% common salt or 3-5% potassium permanganate. This treatment, lasting for a duration of 15 to 30 seconds, has been found to mitigate the transmission of bacterial infections between different ponds effectively.

Fig. 3: Paddy-based Integrated Farming System

Poultry, goat, sheep droppings, cattle manure, and vermicompost serve as fertilizers for the cultivation of natural phytoplankton and zooplankton. The fish consume the plankton. Typically, the need for supplementary feed is minimal within an integrated farming system. Nevertheless, it is crucial to provide supplementary feeding, such as farm-made feeds or pelleted feed, to enhance productivity and maximise profits. This practice should continue until the fish reach a desirable size suitable for consumption, typically weighing between 700 and 1000 grams. According to Reddy *et al*. (2019), regular monitoring of water quality management during culture practices has been shown to enhance the overall success of culture operations. The subsequent accumulation of animal excreta and fish feces has a significant impact on water quality, as well as impeding the growth of fish and making them more susceptible to diseases. The implementation of regular fish sampling at consistent intervals enables the assessment of fish health and growth dynamics. An increased stocking density has been found to be associated with a higher likelihood of disease occurrence. In this particular scenario, the resolution of the issues at hand can be achieved through the substitution of water with freshwater. The survival of fish is contingent upon the presence of dissolved oxygen (DO). An increase in stocking density, phytoplankton, and aquatic weeds has been observed to result in a reduction in dissolved oxygen (DO) levels. As a result, during the pre-dawn period, fish ascend to the water's surface and endeavour to occupy that space. Aeration is necessary during this period. Aeration in the pond can be achieved through various methods, including using a paddle wheel aerator, employing a motor to create a downward-to-upward flow, utilising a boat motor to induce disturbance, and employing oxygen tablets containing chemicals such as calcium oxide.

Partial harvesting can be conducted at the 6-month mark for zero-point culture, the 8-month mark for stunted fingerlings, and the 10-month mark for fingerlings stocked in the pond. The periodic harvesting of fish is conducted in response to demand, employing methods such as cast nets or drag nets. The attainment of marketable size by fish and the subsequent harvesting of fish at this size also contribute to a reduction in stocking density and pressure within the pond. The reintroduction of species into their natural habitat after harvesting activities serves to preserve and sustain the ecological equilibrium. The catla species exhibited the highest growth rate, followed by rohu, mrigal, and common carp. The catla fish reached its maximum size of approximately 3 kilograms at the final harvest. This phenomenon could be attributed to the ingestion of plankton and feed. Fish production can be achieved at a rate ranging from 5 to 12 tonnes per hectare per year. Fish production is contingent upon both the stocking density and the provision of supplementary feeding. According to Reddy *et al.* (2019), the net income derived from fish culture ranges from ₹ 1,20,000 to ₹2,50,000/-, contingent upon the specific management practices employed by the farmers.

Integrating aquaculture and rice production has proven to be a successful strategy in enhancing water-use efficiency as a managerial approach to address the growing water scarcity caused by climate change. According to Miao (2010), reducing competition for water and other resources through diversification of income and food sources can offer a limited safeguard against climate variability. Additionally, the shade created by the rice plants maintains a lower water temperature, thereby creating a more favourable environment for fish production. The temperature of the water is a crucial factor that substantially impacts the growth and overall well-being of fish. Elevated water temperatures can profoundly impact the well-being of fish, manifesting in heightened occurrences of stress and disease, and potentially culminating in large-scale mortality incidents. In addition, elevated temperatures have the potential to reduce the concentration of dissolved oxygen within aquatic environments. Elevated temperatures can also reduce the feed conversion efficiency of fish, necessitating a greater quantity of feed to achieve equivalent fish production levels. In order to mitigate the impacts of rising temperatures, farmers have the option to employ fish refuges and construct ditches along the periphery of the pond. These aquatic refuges provide fish with a habitat characterised by lower temperatures and greater depth, enabling them to seek shelter during peak daytime heat. Simultaneously, it is imperative to select rice cultivars and fish species that demonstrate strong adaptability to the specific environmental conditions of the locality.

Fig. 4: Display of Harvested Fish in the Tribal Area

Cattle Farming

The predominant agricultural system in various regions of India is characterised by a combination of livestock rearing and crop cultivation. Integrating crop cultivation and livestock rearing within the same agricultural operations enables small-scale farmers to diversify income and generate employment opportunities. Crops and livestock exhibit a mutually beneficial relationship, complementing each other. According to Jayanthi *et al.* (2000), farm waste products frequently sustain the animal component of the livestock and crop system. This animal is utilised for land cultivation purposes and also serves as a source of manure, which can be utilised as both fertiliser and fuel. According to Venkatadri *et al.* (2008), a study found that a significant majority of farmers, specifically 98%, expressed the belief that engaging in livestock rearing helps mitigate vulnerability during drought periods. Additionally, 97% of the respondents in the sample reported that farmers in regions with a focus on milk production exhibited lower rates of suicidal tendencies. Conversely, the study also revealed that commercial agriculture in Andhra Pradesh was associated with a higher suicide rate, with 96% of the respondents acknowledging this correlation. Dung encompasses both macro and micro nutrients essential for enhancing soil fertility and promoting increased crop growth. Cow dung is utilised in the biogas production process. Biogas has been identified as a viable option for renewable, alternative, and environmentally benign energy sources (Godi *et al.*, 2013). According to Alam *et al.* (2000), the quantity of manure generated by a pair of crop-integrated draught cattle was reported to be 12 tonnes. A tonne of manure contains 8 kg of nitrogen, 4 kg of phosphorus, and 16 kg of potassium. The application of manure enhances soil fertility and improves the capacity of the soil to retain precipitation. According to Kochewad *et al.* (2017), the presence of grazing animals in plantation areas reduces the expenses associated with weed control. According to the research conducted by Jayanthi (2018), it was observed that the conventional cropping scheme yielded a net return of ₹8422/-, while the integrated farming system (IFS) yielded a net return of ₹10,913/-. Furthermore, the net return for the milk company was found to be ₹8987/-. According to Wanapat *et al.* (2018), the quantity and consistency of feed provided to the dairy unit are crucial factors for ensuring the profitability of a dairy company. According to Nirmala *et al.* (2018), the researchers observed the average milk output in graded Murrah buffaloes across three different feeding conditions. In care I, the buffaloes were provided with paddy straw and natural pasture *ad libitum*. In care II, the buffaloes were fed with deoiled rice bran, with a daily intake of 2 kg for maintenance and 1 kg for every 2 kg of milk production, in addition to paddy straw and natural pasture *ad libitum*. In care III, the buffaloes were given a concentrate feed mixture, with a daily intake of 2 kg for maintenance and 1 kg for every 2 kg of milk production, along with paddy straw and natural pasture *ad libitum*. The observed response to an augmentation in milk production can be ascribed to the elevated protein concentration found in the enriched feed provided to the utilised livestock. According to the findings of Kochewad *et al.* (2017), a pond with an area of 1 hectare can be adequately sustained by a population of 5-6 mature cattle. In conjunction with a quantity of 9000 kg of milk, the farming approach that combines livestock, crops, and fish is projected to produce an estimated range of 3000-4000 kg of fish per hectare per year. The implementation of this machine would result in a reduction of labour required for the lifting of cow dung. The adult cow's dietary needs consist of 9-10 tonnes of green fodder and 2-2.2 tonnes of dry fodder, which can be obtained from the crop portion. The utilisation of compost has the potential to enhance the soil fertility within the pond ecosystem. According to the findings of Ponnusamy and Devi (2017), the integrated systems of Crop + Dairy + Poultry + Fishery, Crop + Dairy + Poultry + Horticulture, and Crop + Dairy + Poultry + Sheep were observed. Goat husbandry and horticulture have been identified as factors that positively impact the overall financial gains of agricultural households. This is particularly relevant for farm families who prioritise profit-driven agricultural enterprises, such as fisheries, fruit cultivation, floriculture, and sugarcane production. The farmers in the study area could generate a substantial income from their small or medium-sized land and livestock holdings due to their diligent management practices, which involved the utilisation of family labour.

Fig. 5: Ongole Cattle maintained by the Small and Medium Farmers

Duck Rearing

The mutually advantageous association between fish and ducks is evident as fish ponds provide ducks with an optimal, disease-free environment. Ducks engage in the consumption of various organisms, including snails, juvenile frogs, and insects. The excrement of ducks is directly deposited into the water, thereby providing the necessary nutrients for the proliferation of plankton. The duck submerges its head fully into the deeper zone, disrupting the sand as it forages for sustenance. This process aids in blending and producing organic food while also serving as a natural means of aerating to enhance the concentration of dissolved oxygen, a vital requirement for the sustenance of fish and microbial communities. Ducks exhibit a feed wastage rate of approximately 10-20% with regards to the provided feed. The aquatic organisms will ingest the contents of this feed. Therefore, the efficient utilisation of feed waste can be utilised. A density of 4 ducks per square meter allows for the rearing of 500 ducklings per hectare of pond. The implementation of partial shelter and fencing can facilitate the provision of a dry walkway for ducks. According to estimates, each duck produces approximately 20 to 25 kg of excrement.

While layer ducks have traditionally been the primary focus of rearing, meat and ornamental ducks have recently been added to the industry. The Khaki Campbell and the Indian Runner are widely recognised breeds of domestic ducks known for their prolific egg-laying capabilities. The Khaki Campbell breed is widely regarded as suitable for rearing in various locations due to its ease of care, robust nature, and ability to freely range. These ducks typically begin laying eggs within three months, provided they are in good health, receive proper nutrition, and are exposed to favourable environmental conditions. Khaki Campbell ducks' laying period typically spans 2 to 3 years. The majority of ducks commence egg-laying once they reach the age of six months. Ducks do not typically lay eggs while in the water and tend to lay eggs primarily during night time. Maintaining adequate shelter and ensuring the presence of sufficient paddy straw are essential for the proper collection of eggs without causing any damage.

This study aims to examine meat consumption patterns, specifically focusing on three duck varieties: Muscovy, Mule, and Aylesbury. These particular varieties exhibit rapid growth rates and demonstrate high efficiency in converting feed into usable energy. The Muscovy variety is well-suited for backyard purposes, and its meat is highly regarded for its exceptional flavour. Males typically exhibit an average weight range of approximately 4 to 5 kg, while females tend to be smaller, with weights ranging from 2 to 3 kg. The Mule variety is derived from a hybridisation between the male Muscovy and the female Pekin. The growth rate of this particular variety surpasses that of Muscovy, resulting in a substantial yield. The rearing of ducks results in a supplementary revenue of ₹8,886/-. The duck-cum-fish community, implemented as a component of the mixed farming scheme, yielded a net benefit of ₹1,441/- from an area measuring 0.027 hectares. According to Ganesan (1989), the mixed farming system yielded a net additional revenue of ₹10,327/- compared to the current cropping system.

Fig. 6: Distribution of Ducks to the Farmers for Encouraging Duck Farming

Poultry Rearing

Poultry rearing is a widely recognised agricultural practice, with many rural households maintaining sufficient birds in their backyard. The selection of meat or egg varieties depends on specific requirements and purposes. The production and productivity of eggs and meat are influenced by various factors such as genetic diversity, nutritional intake, animal health, and environmental conditions. The construction of the poultry shed can be situated either on the pond dyke or on the adjacent side of the dyke, where the excrement can be deposited directly into the water or collected using appropriate sheets. Housing can be constructed utilising either locally available materials or permanent structures. Each bird requires an approximate area of 1 square meter during its initial stages, which increases to 2 square meters as it progresses in development.

The poultry bird varieties, namely Vanaraja, Srinidhi, Rajasri, and Kuroiler, are considered dual-purpose birds suitable for meat and egg production. On the other hand, Gramapriya and BV-380 are particularly well-suited for egg production. It is anticipated that an estimated 500 to 600 avian species will inhabit a water distribution area spanning one hectare. Following the administration of vaccinations against infectious diseases and the implementation of necessary prophylactic measures to safeguard against epidemics, chicks that are approximately eight weeks old are accommodated in poultry houses located in close proximity to the pond. In the Intense Rearing Method, birds are provided with grower mash from the age of 9 to 20 weeks, at a daily rate of 50 to 70 grams per bird. In contrast, avian species that have reached a minimum age of 20 weeks are provided with a daily ration of layer mash, typically ranging from 80 to 120 grams per bird. According to Nirmala *et al* (2017b), the avian species were released in the vicinity of the pond and fields for the purpose of scavenging, thereby meeting their dietary needs within the context of a free-range rearing program. During scavenging, birds typically consume a variety of food sources, including kitchen waste, earthworms, grasshoppers, ants, green grass, leafy vegetables, and seeds. Furthermore, in conjunction with scavenging, all participants provided their avian companions with a quantity of fragmented rice, bajra, sorghum, corn, and other similar grains. Both avian species are provided with an ample provision of water continuously. Egg production typically begins at approximately 22 weeks of age, followed by a gradual decline thereafter.

Fig. 7: Backyard Poultry Entrepreneurs

The annual egg production varied between 140 and 200, depending on factors such as breed, feeding regimen, health status, and environmental conditions in which the avian population resided. The avian specimens are terminated at the age of 18 months. According to the findings of Karunasree *et al.* (2017), the average body weights of Vanaraja, Gramapriya, and Aseel chickens at 48 weeks of age were recorded as 3.22 kg, 2.5 kg, and 1.5 kg, respectively. Additionally, the study indicated that the egg production of Gramapriya (178) was higher compared to Vanaraja (160) and Aseel (62) chickens. Applying poultry manure as a fertiliser leads to an annual yield of 3000-4000 kg of fish, 60,000-100,000 eggs, and over 3,500 kg of meat per hectare of pond. By employing a stocking density of 20,000 fingerlings per hectare for tilapia, common carp, and murrels, along with a chick density of 4,000 per hectare, it is possible to attain a fish production rate of 10 tonnes per hectare.

Fig. 8: Displaying of poultry eggs by the women entrepreneur

Fig. 9: Backyard poultry birds while feeding

Small Ruminants

Sheep and goat husbandry are recognised as lucrative income-generating activities for rural youth and households. In addition to the utilisation of meat and milk, farmers frequently employ the excrement of livestock as a nutrient-rich form of organic fertiliser. Every mature goat produces approximately 750 grams per day of excrement, which is abundant in organic biomass, nitrogen, phosphorus, potassium, and protein. The urine of goats exhibits comparable levels of potash and nitrogen. Using goat droppings in fish ponds is advantageous due to their physical characteristics. These droppings typically consist of 1.0 cm pellets that are coated in a semi-dried state with mucus, allowing them to float upon application. The utilisation of goat waste in the recycling process has positively impacted the growth and overall health of Rohu and Mrigal fish. This method has the potential to yield an annual fish production of approximately 3.5 tonnes per hectare without the need for additional feed or chemical fertilisers.

Small ruminants are effectively integrated into the farming systems of small and marginal farmers in India (Nirmala *et al.*, 2017). The integration of these industries is expected to enhance overall operational efficiency and preserve ecological equilibrium and economic resilience. In a study conducted by Ramana *et al.* (2000), it was observed that lambs and children grazing on silvi-pasture exhibited a daily body weight gain of 54.8 and 36.8 g/head, respectively. In comparison, those grazing on natural grassland showed a weight gain of 41.2 and 26.4 g/head over 478 grazing days. According to Nirmala *et al.* (2017), a study found that 86.66 percent of individuals who keep goats implemented a comprehensive grazing scheme, allowing their goats to graze on community land. Additionally, 85.30 percent of respondents reported providing housing facilities primarily at night for their goats. Furthermore, the study observed a net income of ₹25,400 per hectare per year in a sample where goat rearing (5+1) was combined with crop activity. According to Kuppusamy (1989), the traditional cropping scheme yielded an additional income of ₹11,932 per hectare per year. According to Dawood *et al.* (1996), the crop portion accounted for 66% of the total revenue generated by the IFS, while the goat segment contributed 17% of the revenue. The cost-benefit ratios for crops, dairy, and goat units were 2.2, 1.9, and 2.3, respectively. This observation provides clear evidence of the potential for goat rearing to serve as a supplementary venture, enabling small-scale farmers in the Cauvery delta district to generate additional economic benefits. Hence, the most suitable companies to complement crop components for farmers in the Cauvery delta region would facilitate consistent revenue generation, create additional employment opportunities, and promote effective waste recycling, particularly in the dairy and goat sectors.

Fig. 10: Goat Flock Maintained by the Entrepreneur

Pig Farming

Integrating swine farming or pig farming into integrated farming systems is possible by strategically locating animal housing units on bank embankments. This arrangement facilitates washing waste and cleaning effluents into nearby ponds, allowing for a productive combination of pig raising and fish culture. In instances where an algal bloom or other irregularity occurs, it is common practice to implement a drainage channel to divert the excreta away from the ponds. To mitigate oxygen depletion, the effluent from pigsties is discharged into the pond during daylight hours. The construction of pigsties can be accomplished using affordable materials, although it is essential to ensure that the flooring is composed of cement and sloped towards the pond. Each pig is allocated a floor space ranging from 1 to 1.5 square meters. The required number of pigs would be contingent upon the size of the pond area. A pond with an area of 1000 m^2 is sufficient to accommodate the excrement produced by three pigs. According to the information, breeding three pigs on a pond with an area of 0.1 hectares is possible. According to Tripathi and Sharma (2005), raising two batches of pigs and one batch of fish is possible. This is because pigs typically reach the size suitable for slaughter within 5-6 months, while raising Indian exotic carp fish usually takes place over a duration of 10-12 months.

The onset of puberty in young female pigs typically occurs around eight months, with the first farrowing event typically taking place at approximately one year. On average, the gestation period spans a duration of 114 days. Pigs typically reach their final retail condition at approximately six months of age. Farmers may strategically organise pig breeding operations to facilitate farrowing, taking advantage of the pigs' short gestation period provided favourable conditions such as temperature and housing are met. Under ideal farming conditions, a proficient sow can produce two litters annually and nurture seven or more piglets. Once boars reach the age of 7 months, they can reproduce and be separated and paired with sows at a ratio of 1 boar to 15 sows. If boars can maintain a rate of 3 sows per two days, they will be granted permission to mate. Therefore, it is possible to initiate pig farming with a small initial stock of one or two sows and gradually expand it into a substantial enterprise within approximately one year. Pigs can be bred in both large and small quantities. The business demonstrates versatility by requiring reduced expenses for accommodation and manpower and by operating within a flexible timeframe spanning from the weaning period (8 weeks) to the marketing stage (6 months).

By providing a diet consisting of 3-3.5 kg of feed, the pig exhibits a weight gain of 1 kg per unit of body weight. Young animals can achieve a more cost-effective increase in body weight. Pigs serve as valuable sources of waste materials, including offal derived from slaughterhouses, refuse, waste generated by bakeries, and molasses, among other examples. It is widely acknowledged that pig manure possesses a digestible fish food content of approximately 70 percent, rendering it highly compatible with integrated fish farming practices. According to Yejin *et al.* (1987), fish exhibit a direct feeding behaviour on manure detritus and by-products of nutrient release within the system. Consequently, it can be inferred that fresh pig manure has a greater propensity to enhance fish growth faster than fermented pig manure. The moisture content of pig dung ranges from 69 to 71 percent, while the nitrogen content ranges from 1.36 to 2 percent, and the phosphate content ranges from 0.36 to 0.39 percent. Pigs are fed a diet consisting of pig mash, which contains approximately 16 to 17 percent protein.

Azolla

Azolla, a hydrophytic fern, is frequently employed as a bio-fertiliser in the cultivation of paddy crops. The fern forms a symbiotic association with blue-green algae, specifically Anabaena azollae, which play a crucial role in nitrogen fixation. Among the diverse species of the genus *Azolla, Azolla pinnata* holds a prominent position. *Azolla* is considered a valuable feed supplement for animals, poultry, and fish due to its elevated crude protein content, surpassing 20 percent on a dry matter basis. Also, *Azolla* contains essential amino acids, particularly lysine, vitamins A and B, and minerals including calcium, phosphorus, potassium, and magnesium (Nirmala *et al.*, 2017c). Photosynthesis necessitates the presence of light and exhibits optimal growth in environments characterised by partial shade. *Azolla* typically necessitates an optimal range of 25 to 50 percent of total sunlight exposure to facilitate its regular growth and development. The growth and multiplication of *Azolla* are contingent upon the presence of water, making it highly vulnerable to water scarcity. It is imperative to uphold an optimal water level within the pond, ensuring it remains at a minimum depth of 4 inches. Organisms exhibit variations in their temperature requirements, as dictated by their respective optimum temperature specifications. The optimal temperature range is between 20 to 30°C. The temperature exceeding 37°C would have a significant impact on the proliferation of *Azolla* through the process of multiplication.

The incorporation of *Azolla* cultivation in rice fields is expected to have a positive impact on piscivorous species. Fish farming in rice fields has been found to have several beneficial effects on the soil. One such effect is the loosening of the soil, which promotes aeration. Additionally, fish farming in rice fields has been observed to enhance the decomposition of organic matter and facilitate the release of nutrients from the soil. The presence of fish excreta in rice fields serves as a direct source of fertilisation for the water, leading to an increase in the available nitrogen supply for the rice crop. The integration of allied components, such as *Azolla* and fish, with rice cultivation in lowland farming has the potential to enhance the recycling of bioresources, as Jayanthi (2018) suggested. Furthermore, implementing green leaf manure, specifically the rice-azolla + fish farming approach with 75 percent recommended nitrogen application, increased agricultural output, improved financial gains, and enhanced soil quality by recycling organic residues. In the rice-azolla + fish farming system, incorporating *Sesbania rostrata*, there was more organic residue addition and nitrogen input through recycling. According to Balusamy (1996), the nutritional value of the unused fish meal, decayed azolla, and fish excreta found at the bottom of the fish trench was significantly higher. These organic materials have the potential to be recycled and utilised for soil enrichment purposes.

According to the findings of Ponnusamy and Devi (2017), the agricultural systems involving the combination of crop, dairy, poultry, fishery, horticulture, sheep and goat, and other related activities were observed to yield a greater net income for farm families. This can be attributed to their involvement in profit-driven farming enterprises such as fisheries, vegetable cultivation, flower production, and sugarcane cultivation. The farmers in the study area could generate a substantial income from their small or medium-sized holdings and small livestock operations, primarily due to their diligent management practices, which included the utilisation of family labour.

Fig. 11: Scientist Explaining About the Role of Azolla to the Farming Community

Beekeeping

Beekeeping is an agricultural endeavour that allows farmers to supplement their income. Although the practice of gathering honey from forests has been prevalent for an extended duration, honey beekeeping has emerged as a means of generating income and achieving economic stability, particularly for individuals with limited land for agricultural purposes. In contemporary times, the growing awareness of health-related issues has led to a notable rise in the market potential for honey and its associated products. Consequently, this trend has brought attention to the emergence of beekeeping as a viable entrepreneurial pursuit. Honey and wax are regarded as economically viable products within honey-based commodities.

Beekeeping is considered a significant agro-based industry that distinguishes itself from other agricultural sectors by not relying on raw materials sourced from artisans. Nectar and pollen, which are readily abundant in the natural environment, serve as the primary raw materials. Honeybees serve as significant pollinators, collecting nectar from flowers and transforming it into honey, which is then stored within the combs of their hive. Beekeeping can be initiated with a solitary colony. Beekeeping has the potential to serve as a constituent element within an integrated farming system.

When considering the benefits of apiculture, it is evident that it offers time-saving advantages and requires minimal attention. Unlike other agricultural practices or animal husbandry, apiculture demands only a few hours per week to maintain the colonies, without the need for additional resources or inputs.

In addition, establishing apiaries has provided a source of income for rural and tribal families, contributing to their long-term sustainability and enhancing their nutritional security. Honey-based products such as honey, brood, and pollen are widely recognised as valuable protein sources. Honey is a primary component in most Ayurvedic and traditional medicinal practices.

There exist four widely recognised species of honey bees distributed across the globe. Four species of bees are commonly known: the Rock bee (*Apis dorsata*), the Little bee (*Apis florea*), the Indian bee (*Apis cerana*), and the European bee (*Apis mellifera*). Within the group of four bee species under consideration, namely *Apis dorsata* and *Apis florea*, it is noteworthy that these particular species exhibit a wild nature and engage in open nesting behaviour, rendering them unsuitable for domestication within contemporary hive structures. The honey production of *Apis dorsata* colonies ranges from 30 to 40 kg annually, whereas *Apis florea* colonies yield approximately 500g of honey per colony. *Apis cerana* and *Apis mellifera* are commonly kept in contemporary apiaries and exhibit cavity nesting behaviour, meaning they reside within enclosed structures. Hive bees demonstrate the ability to construct multiple parallel combs, whereas wild bees typically construct only a single comb. There is significant variation in the size of both the body and comb cell size among bees.

Apis cerana and *Apis mellifera* exhibit complementary characteristics while also displaying distinct adaptations. *Apis cerana* exhibits greater adaptability to high-altitude regions. *Apis mellifera* exhibits greater success in flat, low-lying regions. The hive comprises several components, including the floorboard, brood chamber, top cover, frames, and entrance rod. These components can be readily disassembled. The hive's structure can vary, with the possibility of it being either double or single-walled. The single-walled hive is characterised by its lightweight construction and affordability. The optimal period for initiating beekeeping in a given area is during the onset of the swarming season. Swarming is an inherent behaviour observed in bees, wherein they divide their colonies under favourable circumstances that typically promote the well-being of both the original colony and the newly formed swarm. This phenomenon takes place in late spring or early summer. The area/IFS model should possess diverse bee forage crops that offer ample quantities of nectar and pollen. The primary sources of nectar and pollen include maize, coconut, cucurbits, guava, citrus, mango, moringa, banana, sunflower, mustard, lucerne, mustard, and sesame.

It is recommended that the bee hives be positioned close to the field, as this would minimise the energy expenditure of bees during migration in their quest for honey. Italian bees, with a density of 3 beehives per hectare, and Indian honey bees, with a density of 5 bee hives per hectare, are intended to be situated near a field where 10% of flowering resources are accessible. It is recommended that each beehive box contain a minimum of 5 to 6 frames, populated by a robust colony of bees led by a young and vigorous queen. By implementing a stocking density of three colonies of *Apis mellifera* per hectare, an annual honey production of approximately 50-60 kg can be achieved. Similarly, by adopting a stocking density of five colonies of *Apis cerana* per hectare, a honey yield of approximately 25-30 kg per year can be obtained. Additionally, it is worth noting that this practice also increases crop yield, providing an additional advantage.

Fig. 12: Apiary Entrepreneurs

Silvi-pasture based Farming System

In this scheme, improved fodder varieties are practiced concurrently with annual trees on the same plot of land. The process necessitates the provision of sustenance to animals through the provision of tree leaves as a form of fodder. This approach addresses the scarcity of green fodder and reduces the financial burden of providing livestock with concentrated feed during periods of low availability. Various improved varieties of fodder, including fodder sorghum, fodder maize, guinea wheat, Napier grass, para grass, stylo, fodder cowpea, berseem, lucerne, hedge, and fodder plants such as *Leucaena leucocephala, Bauhinia variegata, Albizia lebbeck, Albizia amara, Moringa oleifera, Sesbania sesban, Sesbania grandiflora*, and *Hardwickia binata,* are utilised in different regions of the country's silvi-pastoral systems. The utilisation of an integrated agricultural system involving the cultivation of sorghum and cowpea, as well as the combination of *Leucaena leucocephala* and *Cenchrus ciliaris,* along with *Acacia Senegal*, has been observed. Ramasamy *et al*. (2007) found that integrating goats with *Cenchrus ciliaris* increased 113 man-days per hectare per year in employment opportunities in dryland areas.

Ehsanul (2016) states that the mixed crop livestock scheme provides animals with fodder derived from grass, nitrogen-binding legumes, weeds, and seed residues. Animals engage in the practice of grazing either beneath trees or on stubble, thereby facilitating the cultivation of drought-resistant crops and the production of compost, all the while serving as a means of conserving resources. In this context, the cultivation of Napier grass holds significant potential to contribute to Bangladesh's integrated farming practices. The nutrient content of Napier grass is significantly elevated. It is recommended that harvesting be conducted a minimum of three times within a given calendar year. Farmers can potentially increase their profits by engaging in the sale of grass. Napier grass can expeditiously fulfil the roughage requirements of both indigenous and hybrid dairy cattle. Fodder cultivation can be conducted on the terraced embankments surrounding the pond, providing sustenance for grass carps and other animals.

Fig. 13: Farming requesting advice for growing the fodder cultivation

Fig. 14: Fodder harvested for feeding of cattle

Vermicompost

Fertilisation or manuring is a commonly employed practice in fish ponds for the purpose of promoting natural fish production. This practice is considered crucial for ensuring sustainable aquaculture and reducing the financial burden associated with expensive feeds and fertilisers, which account for over 50 percent of the total input cost (Oribhabor and Ansa, 2006; Chakrabarty *et al.*, 2009). Vermicompost is a highly valuable biofertiliser and soil compost due to its superior nutrient content. According to Chakrabarty (2008), the vermicompost treatment resulted in higher average weight and overall fish yield than the treatments involving mixed fertiliser and single super phosphate. The vermicompost used in the study contained remnants of the body and earthworm coconut, which is rich in iron, protein, growth hormones, and other substances that contribute to fish production. According to Ghosh (2004), fish subjected to vermicompost care exhibited an average weight gain of 1.64 g per individual per day when maintained at 15,000 fish per hectare density.

Fig. 15: Scientist Explaining about the Importance of Vermicompost to the Visitors

Conclusion

Small and marginal farmers require financial support from diverse sources to sustain their operations and achieve a twofold increase in their earnings. In contrast to monocropping, multiple cropping with diverse cropping systems consistently results in higher income generation. High-value crops typically offer farmers advantageous prospects for generating supplementary income. In contrast, the IFS consistently prioritise enhancing production and productivity, aiming to optimise the utilisation of available resources and market linkages.

References

Alam, M.R., Sarkar, R.I., Hossain M.D. and Islam, M.S. (2000). Contribution of livestock to small farms in Bangladesh. Asian Australian Journal of Animal Science, 13: 339-342.

Anon (2010). A manual on integrated farming systems (IFS). Caribbean Agricultural Research and Development Institute (CARDI), pp. 58.

Aquino, D., Barrio, A.D., Trach, N.X., Hai, N.T., Khang, D.N., Toan, N.T. and Hung. N.V. (2020). Rice Straw-Based Fodder for Ruminants. M. Gummert et al. (eds.), Sustainable Rice Straw Management, pg. 111- 129.

Balusamy, M. (1996). Studies on nitrogen management in lowland rice-fish-azolla integrated farming system. Ph.D. Thesis, TNAU, Coimbatore.

Bhardwaj, R.L., Bajpai, N.K. and Kumar, D. (2014). Bench mark survey on progress in vegetable cultivation in district Sirohi. Krishi Vigyan Kendra, Sirohi, Agriculture University, Jodhpur, pp. 12–15.

Bhardwaj, R.L. and Nandal, U. (2015). Papaya cultivation brought socio-economic and nutritional security in Sirohi tribals of Rajasthan, India: A case study. Asian Journal Dairy and Food Research, 34(1): 44-48.

Chakrabarti, A., Amitava, Dey, and Dharmendra, K. (2014). Livestock cum Fishery integrated farming system. Krishisewa.

Chakrabarty, D. (2008). Vermicompost and organic pisciculture. Akshay Krishi Vikash (Publisher), West Bengal, India: 1–16.

Chakrabarty, D., Das, S.K., Das, M.K. and Bag. M.P. (2009). Assessment of Vermicompost as direct application manure in fish farming ponds. Turkish Journal of Fisheries and Aquatic Sciences, 10: 47-52.

Dawood, S.P., Santhi, A., Ponnsamy, A. and Muthukrishnan. P. (1996). Integrated Farming System for lowland and Cauvery delta zone. Farming System. 13: (3-4): 11-14.

Dawson, T.E. (1996). Determining water use by trees and forests from isotopic, energy balance and transpiration analyses: the roles of tree size and hydraulic lift. Tree Physiology. 16: 263-272.

Ehsanul H.M.D. (2016). Suitable Integrated Crop- Livestock Production System in Char area of Bangladesh. Journal of Fisheries and Livestock Production. 4(1):156.

Ganesan, S. (1989). Studies on the feasibility of introducing Duck cum fish rearing in the deltaic region of Thanjavur District. M.Sc. (Ag.) (diss.) Tamil Nadu Agricultural University, Coimbatore.

Ghosh C. (2004). Integrated vermin-pisciculture- an alternative option for recycling of solid municipal waste in rural India. Bioresource Technology. 93: 71-75.

Godi, N.Y., Zhengwuvi, L.B., Abdulkadir, S. and Kamtu, P. (2013). Effect of cow dung variety on biogas production. Journal of Mechanical Engineering Research. 5:1-4.

Gururaj B., Hamsa K.R., Ramesh, and Mahadevaiah G.S. (2017). Doubling of Small and Marginal Farmers Income through Rural Non-Farm and Farm Sector in Karnataka. Economic Affairs. 62(4): 581-587.

Horton, J. and Hart, S.C. (1998). Hydraulic lift: A potentially important ecosystem process. Trends in Ecology & Evolution. 13(6):232-235.

IPCC, (2007). Solomon, S., Qin, D., Manning, M., Chen, Z., Marquis, M., Averyt, K. B., Tignor, M. and Miller, H. L. (eds.), "Climate Change 2007: The physical science basis", Contributions of working group I to the Fourth Assessment Report of the IPCC, Cambridge University Press, Cambridge.

Jayanthi, C. (2018). Integrated rice based farming system for enhancing productivity and climate resilience. Oryza. 55 Special issue: 64-70.

Jayanthi, C., Rangasamy, A. and C. Chinnusamy. 2000. Water budgeting for components in low land integrated farming systems. Agricultural Journal. 87:411-416.

Karunasree, E., Nirmala, T.V. and Reddy, R.V.S.K. (2017). Comparative performance of different varieties of chicken under backyard system of rearing in tribal and rural areas of West Godavari district, AP. International Journal of Science, Environment and Technology. 6(2): 1100-1104.

Kashyap, Poonam, Kansal, A., Prusty, A.K. and Singh. J.P. (2015). Evaluation of horticulture based IFS models for providing nutritional security to small and marginal farmers of western plain zone of Uttar Pradesh, India. International Journal of Plant Ecology. 2(1):15-7.

Kochewad, S.A., Meena, L.R., Kumar, D., Kumar, S., Meena, L.K., Singh, S.P. and Khushyal Singh. (2017). Livestock based integrated farming systems for livelihood improvement of small and marginal farmers. South Asian Journal of Food Technology and Environment, 3(1): 526-532.

Kuppusamy, M. (1989). Feasibility of introducing sericulture as a component in the farming system in Thanjavur district. M.Sc. (Ag.). Thesis, Tamil Nadu Agricultural University, Coimbatore.

Miao, W.M. (2010). Recent developments in rice-fish culture in China: a holistic approach for livelihood improvement in rural areas. In S.S. De Silva & F.B. Davy, eds. Success stories in Asian aquaculture. Dordrecht, Heidelberg, London & New York, Springer. pp. 15-40.

Nirmala, T.V, Devivaraprasad Reddy, A., Karunasree, E., Reddy, R.V.S.K., Venkata Subbaiah, K., Shali Raju, G. and V. Deepthi. (2018). Effect of concentrate feed mixture on yield of milk in graded Murrah buffaloes. Multilogic in science. 8: 6-7.

Nirmala, T.V., Karunasree, E., Reddy, A.D., Reddy, R.V.S.K., Subbaiah, K.V, Raju G.S. and V. Deepthi. (2017). Adoption of scientific management practices in goat farming by tribal goat farmers in West Godavari district of Andhra Pradesh. Journal of Pharmacognosy and Phytochemistry. SPI: 536-539.

Nirmala, T.V., Karunasree, E., Reddy, R.V.S.K., Reddy, A.D., Subbaiah, K.V., Raju, G.S. and J.V. Prasad. (2017). Importance of Azolla as feed supplement for livestock and poultry. Bioquest. (1): 29-30.

Nirmala, T.V., Prasad, S., Subrahmanyeswari, B. and K.N. Kumari. (2017). Status of Indigenous chicken rearing in Krishna district of Andhra Pradesh. The Indian Journal of Veterinary Science and Biotechnology. 12(3): 52-55.

Oribhabor, B.J. and Ansa, E.J. (2006). Organic waste reclamation, recycling and re-use in integrated fish farming in the Niger Delta. Journal of Applied Sciences and Environmental Management, 10(3): 47–53.

Ponnusamy, K. and Devi. M. K. (2017). Impact of Integrated Farming System Approach on Doubling Farmers' Income. Agricultural Economics Research Review.30: 233-240.

Ramakrishna, R., Shipton, T.A. and Hasan, M.R. (2013). Feeding and feed management of Indian major carps in Andhra Pradesh, India. FAO Fisheries and Aquaculture Technical Paper No. 578. Rome, FAO. 90 pp.

Ramana, D.B.V., Rai, P., Solanki, K.R. and Singh. U.P. (2000). Comparative performance of lambs and kids under silvi-pastoral system In: Proc. III Biennial ANA conference-2000, Hisar, pp. 47-48.

Ramasamy, C., Natarajan, S., Jayanthi, C. and Kumar, D.S. (2007). Intensive integrated farming system to boost income of farmers. In: Proceedings of 32nd IAUA vice chancellors annual convention on diversification in Indian Agriculture, Birsa Agricultural University, December 20-21, pp. 28-47.

Reddy, A.D., Nirmala, T.V., Karunasree, E., Subbaiah, K.V., Raju, G.S., Deepthi, V., Reddy, R.V.S.K., Srinivasulu, B., Prasad, J.V. and Prasad. Y.G. (2019). Diversification in agriculture through fisheries based integrated farming system for doubling farmers' income in agency areas of West Godavari district of Andhra Pradesh- A case study. International Journal of Chemical Studies; SP6: 930-932.

Reddy, R.V.S.K., Karunasree, E., Reddy, A.D., Deepthi, V., Nirmala, T.V., Raju G.S., Subbaiah KV and Prasad. J.V. (2018). An evaluation study on viable Integrated Farming System (IFS) model in Godavari delta of Andhra Pradesh. Journal of Pharmacognosy and Phytochemistry. SPI: 2361-2368.

Shyam, R. and K. Kumar. (2001). Fish-Livestock Integrated Farming. Central Institute of Freshwater Aquaculture, Bhubaneswar, Orissa. 68pg.

Soni, R.S., Katoch, M. and Ladohia. R. (2014). Integrated Farming Systems - A Review. IOSR Journal of Agriculture and Veterinary Science. 7(10): 36-40.

Tripathi, S.D. and B.K. Sharma. (2005). Integrated fish-pig farming in India. Integrated Agriculture-Aquaculture: A Primer. Animal-Fish Systems.

Venkatadri, S., Swaroopa Rani, K. and G. Raghunadha Reddy. (2008). A study on improvement in rural livelihoods through dairy farming. Centre for Self Employment and Rural Enterprises. National Institute of Rural Development, Rajendranagar, Hyderabad

Wanapat, M., Foiklang, S., Sukjai, S., Tamkhonburi, P., Gunun, N., Gunun, P., Phesatcha, K., Norrapoke, T. and Sungchhang, K. (2018). Feeding tropical dairy cattle with local protein and energy sources for sustainable production. Journal of Applied Animal Research.46:1, 232-236.

Yejin, Y., Yun, Z., Dan, H. and W. Junhua. (1987). Preliminary studies on the effects of fresh and fermented pig manures on fish production. Proceedings of integrated fish farming systems. Freshwater Fisheries Research Centre of Chinese Academy Fishery of sciences, pp.:324-327.

17

Integrated Farming System The Road to Sustenance

***Longjam Boris Singh*[1*]*, S. Zeshmarani*[1]*, Yanglem Herojit Singh*[2]**
***Chuwang Hijam*[1] *and S. Prabin*[1]**

[1]*Krishi Vigyan Kendra, Thoubal, Manipur*
[2]*Central Agricultural University, Imphal, Manipur*

Abstract

The socio-economic sustainability of marginal and small farmers is at risk due to the shrinking agricultural land, rising population, and alterations in microclimate, posing a challenging task for them. The integrated farming system (IFS) offers a solution for small and marginal farmers in rural areas to meet the growing demands of socio-economic activities. It involves the combination of crop and livestock enterprises, which are separate but interconnected components that work together harmoniously. Maximisation, rejuvenation, pests and diseases avoidance and minimal use of chemical inputs are the main goals of integrated farming system. The rationale behind integrated farming systems is to reduce waste produced by various subsystems on the farm. Due to its holistic approach, integrated farming system may serve as the pathway to doubling farmers' income. Production and productivity can be increased significantly in IFS, through proper resource utilisation and labour management. Low volume of produce and high costs of transport and inputs still pose a serious hurdle to integrated farming system.

Keywords: Mixed farming, holistic approach, crop-livestock, resource management.

Introduction

Farming has undergone significant changes since the Green Revolution in the late 1960s and economic liberalisation in the early 1990s. These changes have led to a shift in focus towards a few specific enterprises due to factors such as a decrease in farm size, fluctuating commodity prices, the need for livelihood diversification, and a shortage of labor. (Ponnusamy and Kousalya, 2017).

The current scenario in the country indicates that the agricultural area under cultivation is decreasing and more than 20 percent of the current cultivable land will be converted into non-agricultural purposes by 2030 (Gill *et al.*, 2005). 80 per cent of farm families belong to the small and marginal farmer categories. Hence India is putting more emphasis on agriculture in research as well as in technology transfer through many extension activities. The vital solution to food and nutrition security may be provided by production enhancement through natural farming in India in general and particularly in the North East. Furthermore, the utilisation of traditional farming methods in crop cultivation has contributed to a decline in productivity rates (Velu, 2011). Nutritional and food security will be provided by production enhancement and constraints in conventional/natural farming such as low productivity by adopting scientific agronomic practices and technologies. Furthermore, the widespread use of chemical fertilisers and pesticides not only degrade the healthy environment but also increases operational costs. This has raised questions regarding sustainability and feasibility in agriculture (IAASTD, 2009).

Concept of Integrated Farming System

Integrated farming system is an agricultural method that aims to enhance productivity in crop cultivation while ensuring efficient resource utilisation. This approach involves the integration of crop and livestock enterprises, which are distinct yet interconnected components of the farming system. (Okigba, 1995). “There is no waste and waste is only a misplaced resource which can become a valuable material for another product in IFS” (FAO, 1997). Integration is a key aspect of the integrated farming system, where the outputs of one enterprise are utilised as inputs by another enterprise within the farming system. This system aims to minimise risks, increase overall production and profits by reducing external inputs through recycling and improving the utilisation of organic wastes and crop residues. The main objective of the integrated farming system is to select a few interdependent, interrelated, and interlinking production systems that focus on specific crops, animals, and related subsidiary activities. Additionally, the effective linkage and complementarity of various components are crucial for developing a holistic approach to the integrated farming system(Rao *et al.*, 1997). Indeed, integrated farming system is an approach that includes different land-based enterprises such as crop, livestock, mushroom, apiary, fishery etc., which are integrated into the biophysical and socioeconomic situations taking farmers' preferences and goals into consideration. All in all, IFS is the solution for small and marginal farmers which involves a multidisciplinary approach.

Rationale of Integrated Farming System

The integration of farming systems aims to minimise waste by utilising by-products and waste from one subsystem as inputs for other subsystems. This approach not only increases productivity but also reduces the costs of outputs (Edwards *et al.* 1988, Gill *et al.* 2009). Due to land fragmentation, scarcity of agricultural inputs and changing climatic scenarios in the country, a holistic perspective in agriculture is necessary. The integrated farming system provides an opportunity for the intensification and diversification of crops and the integration of allied enterprises. It offers the opportunity for recycling nutrients, minimising the dependence on chemical fertilisers for crop production. Employment generation throughout the year is also possible under IFS. In agricultural production systems, diversification with different crops and enterprises helps bring stability and enables farmers to meet the diversified needs of the family. IFS also offers insurance against crop failure and market risks. Moreover, the overall sustainability of the soil is much better compared to the conventional mode of farming (Manjunath and Singh, 2012). The implementation of Integrated Farming Systems (IFS) involves various practices such as raising fish in ponds and rearing pigs, ducks, or chickens in pens. These practices not only increase profitability but also enhance sustainability by utilising animal and feed wastes effectively (Vincke 1988, Gill *et al.,* 2005). In conclusion, the integrated farming system effectively promotes self-sufficiency among farmers by reducing risks and maintaining environmental harmony, ultimately leading to an enhanced standard of living (Mali *et al.*, 2014).

Objectives of Integrated Farming System

1. To identify existing farming system and the relative viability.
2. To formulate models for different farming systems and allied enterprises under different farming situations.
3. To ensure conservation and utilisation of available resources and effective recycling of farm residues and by-products within the system.
4. Maintenance of sustainable production system without damaging the resource base and environment.
5. To increase the profitability of the farm by complementing main and allied activities.

Goals of Integrated Farming System

The four primary goals of IFS are

i. Optimise the output of each individual enterprise to ensure a consistent and reliable income stream.

ii. Enhance the efficiency of the system and achieve a balanced agro-ecological state.

iii. Minimise the accumulation of insect-pests, diseases, and weed populations by implementing effective management strategies within the cropping system. By adopting natural approaches such as proper crop rotation, integrated pest management, and weed suppression techniques, these harmful factors can be controlled and their intensity reduced.

iv. Promote a healthier environment and produce by minimising the reliance on chemicals such as fertilisers and pesticides, thereby ensuring the production of chemical-free and wholesome crops for the benefit of society.

Benefits of Integrated Farming System

i. Profitability: The profitability of enterprises can be significantly boosted by employing the by-products of one enterprise as inputs for another, thereby reducing production costs (Maheswarappa *et al.*, 1998; Manjunath and Itnal, 2003; Ravisankar *et al.*, 2010).

ii. Sustainability: The utilisation of interconnected components in IFS allows for the preservation of the resource base for extended periods, ensuring sustenance. Additionally, this approach helps to minimise the accumulation of pests and diseases, thereby promoting a balanced agro-ecological system (Korikanthimath and Manjunath, 2009; Gill *et al.*, 2010; Kumar *et al.*, 2011).

iii. Environmental safety: The effective recycling of waste materials through the integration of suitable enterprises and components in IFS leads to a reduction in environmental pollution. Additionally, IFS reduces the farm's reliance on external inputs, thereby minimising pollution caused by their excessive use (Shukla *et al.*, 2002).

iv. Recycling of resources: Due to the recycling of wastes and by-products, IFS relies less on outside inputs. Hence, the production system is more stable in IFS (Kumar *et al.*, 2013).

v. Year-round farm income: The production of farm produce, such as milk, eggs, mushrooms, and fruits, is facilitated by different elements of the Integrated Farming System (IFS). This system ensures a consistent and continuous income stream throughout the year (Behera and Mahapatra, 1999; Maheswarappa *et al.*, 2001; Kumar *et al.*, 2013).

vi. Risk minimisation: In contrast to single commodity-based agriculture, which is vulnerable to natural hazards like droughts, floods, and disease outbreaks, the Integrated Farming System (IFS) approach involves diversifying the farm with various enterprises such as intercropping, agroforestry, horticulture, livestock, and plantation crops. By integrating different enterprises, the risk is reduced, and farmers in the area can increase their income. Therefore, the IFS model offers a sustainable production system that is resilient to climate change and minimises risk (Ayyappan and Arunachalam, 2014).

Introduction of Integrated Farming System

The successful implementation of an integrated farming system requires a significant upfront investment, including the establishment of animal sheds and the purchase of livestock. However, the lack of financial resources and the high costs associated with inputs pose a challenge to the adoption of this farming system (Pandey *et al.*, 2019). The adoption of IFS demands proper skill and knowledge of fodder crops, availability of quality planting materials for livestock components as well as proper availability of health services for the livestock (Pushpa, 2010). Hence, the adoption of IFS needs motivation and encouragement through training and demonstration. Moreover, credit facilities and the proper attitude of the farmer are necessary for aptly adopting the IFS model.

Scope of Integrated Farming System

The primary objective of integrated farming systems is to attain profitable outcomes while ensuring consistent and elevated levels of production, while also minimising the adverse effects associated with intensive farming practices (Lal and Miller 1990, Gupta *et al.* 2012). Within the IFS model, there is a strong emphasis on achieving enhanced productivity while simultaneously addressing environmental degradation and improving the living conditions of resource-poor farmers (Kumar *et al.* 2013). A farm, comprising enterprises such as crops, livestock, poultry, fish, and tree crops, is considered as a unit and the enterprises are combined considering factors such as:

- Economics of the proposed integrated farming system.
- Utilisation level of the resources.
- Management skills of the farmers.
- Availability of the resources.
- Soil and climatic factors.

Components of Integrated Farming System

Maximising the complementarities and minimising competition between the enterprises in the model should be the cardinal principle of selection of enterprises in an integrated farming system. At the same time, the selection of a few interdependent, interrelated and inter-linked production systems based on a few crops, animals and related subsidiary enterprises should be the focus of an integrated farming system. The utilisation of primary or secondary produce such as waste from one process, with or without treatment, as raw material for another system makes the entire system mutually interdependent as a whole unit. Since provisions of production such as energy, fertilisers, and feeds are provided, optimum productivity is achieved at minimum costs in an integrated farming system.

Fig 1: Components of IFS

Crop-livestock Component

In the IFS model, natural resource-based conservation, production enhancement through efficient utilisation, optimised allocation and rational management of natural resources are indeed compulsory. Interaction of crops and animals in a synergistic manner has a positive and greater total effect on the yield and soil fertility. Incorporation of small-holder livestock production with less land and commercial livestock production dominated by poultry and dairy is encouraged under an integrated farming system. Soil organic matter can be significantly increased with the application of livestock manures, thus improving soil physical properties and providing NPK and other minerals. Moreover, manure and urine of animals accelerate the decomposition of organic matter and termite activity (Brouwer and Powell 1995, 1998). Crop and animal integrated farming model such as crop and dairy, dairy and vegetables, horticultural crops and dairy are also very popular in the Indian context (Tripathi and Rathi, 2011).

Agroforestry Component

Agroforestry systems offer products such as resins, gums, fibres, oils, medicines, timber and fuel. Arable crops are cultivated in between the perennial trees fruit trees or the multi-purpose tree rows. In case of a reduction in yield due to canopy closure, the system is converted into a horti-silvicultural system. Fodder grasses such as Guinea grass (*Panicum maximum*), Dinanath grass (*Pennisetum pedicellatum*), Napier grass (*Pennisetum purpureum*) and fodder legumes such as stylo (*Stylosanthes hamata*), cowpea (*Vigna sinensis*) are grown in the inter-spaces of perennial fruits or multipurpose tree rows. With suitable management practices, this system can provide a sustainable production system. Due to the permanent grass cover, this system is also very effective for controlling soil erosion (Patra, 2016).

Fisheries-pond Component

Water-based agricultural systems provide an opportunity for effective and efficient recycling of farm nutrients. Feeding animal wastes from the animal houses directly into the pond can be done to feed the fish (Kumar *et al.*, 2018). Fish such as *Labeo rohita* (column feeder), *Catla catla* (surface feeder) and *Cirrhinus mrigala* (bottom feeder) can be raised in a composite manner in IFS. Adult fish can be harvested 3 times at 20-days intervals. Poultry droppings/ litter can be used as fish feed while some of the droppings can be used as crop manure. Water can be drained from the pond where the settled silt can be removed and applied as organic fertiliser to the crops (Kumar *et al.*, 2022). In the paddy-fish mode of integrated farming system, selling fish offers rapid investment returns and creates opportunities for household investment along with an increase in rice productivity. Increased in land productivity, better management of land and crops, and reduced risk of crop failure are additional benefits provided by the integrated farming model (Yassi *et al.*, 2020).

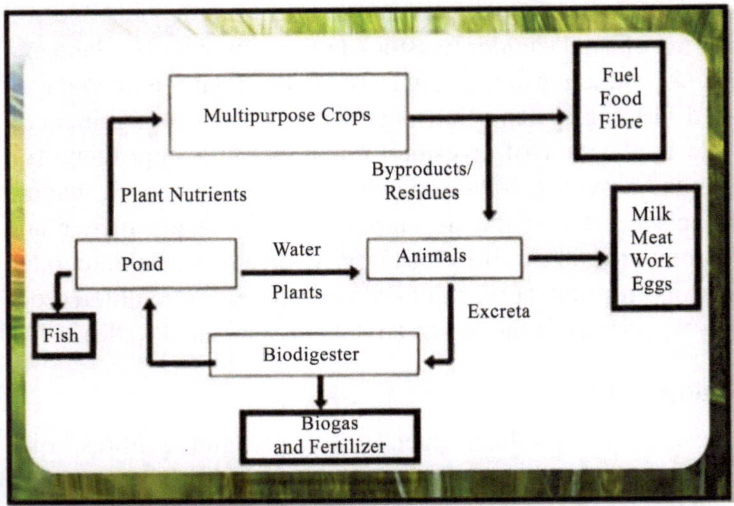

Fig. 2: Schematic Layout of an IFS model

Mushroom Component

The incorporation of a mushroom unit in the IFS model can be done using a small hut made of locally available materials. Paddy straw mushroom (March-September), milky mushroom and oyster mushroom (October to February) can be successfully grown in the unit on raised bamboo racks in the shed. Maintenance of 75-80% humidity is essential by sprinkling water over the walls of the huts and bags (Kumar *et al.*, 2022). Paddy straw is often wasted after paddy harvesting either by burning or left in the field, especially in North Eastern India. This waste paddy straw can be used for mushroom cultivation to enhance food and nutrition security. It has been found that from a total of 4.5t/ha of paddy straw, 2.25t can be utilised for mushroom production, resulting in a yield of 1.15t. Additionally, 0.28t of vermicompost can be obtained from the spent mushroom substrate, which can serve as a nutrient source for paddy and vegetable cultivation in the following year. Additionally, the utilisation of spent mushroom substrate and compost improves soil structure, provides plant nutrition, improves enhances microbial population, increases soil cation exchange capacity, improves plant root structure, regulates pH levels, and enhances soil water status. Therefore, integrating a mushroom unit into the integrated farming system not only improves soil health but also increases the farmer's income.

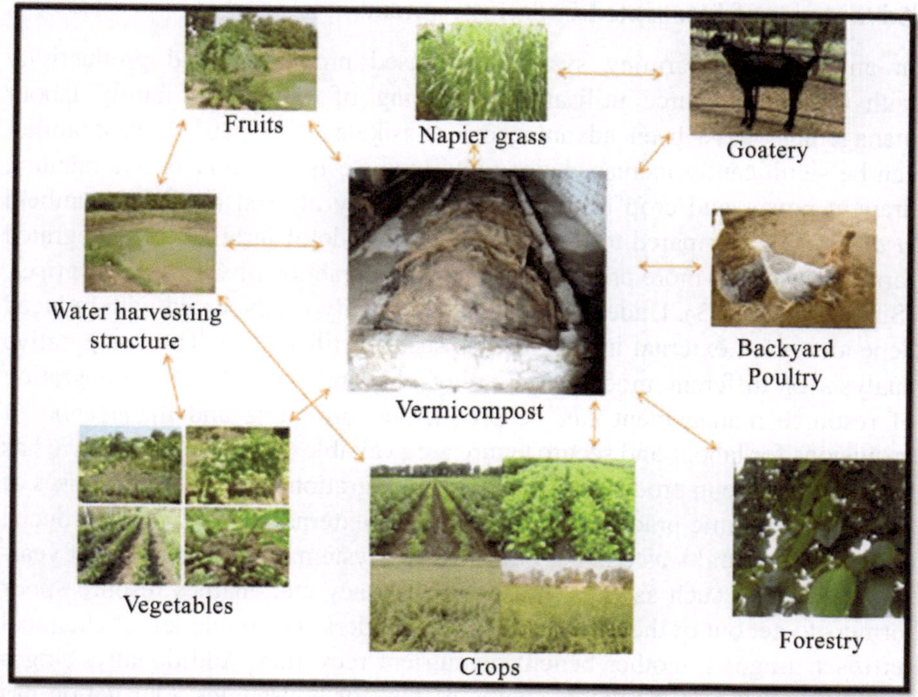

Fig. 3: An example of Integrated Farming System model developed at the ICAR-Central Institute for Cotton Research, Nagpur, Maharashtra.

Role of Integrated Farming System in Doubling of Farmers' Income

Indian farmers face acute problems including the decline in farm income and an increase in the costs of inputs like seeds, fertilisers, herbicides, irrigation and labour. According to the National Institute of Agricultural Economics and Policy Research (NAIP), 70% of the country has an annual income of less than ₹15,000. The lack of infrastructure for growth in most of our states is profound. Hence, the need for doubling farmers' income has increased manifold. About 70 per cent of the farmers are marginal, while 77 per cent of them earn an income of ₹6067 per capita per year. Almost 40 million farmers have a landholding of only 500 m^2, which is far from being sustainable. The Integrated Farming System approach encompasses a comprehensive and interdisciplinary approach that addresses various aspects of farming, such as problem-solving, location-specific considerations, and farmer-centric strategies. This approach plays a crucial role in sustainable development by incorporating economic, logical, and social objectives alongside agricultural food production (Kumar, 2013).

Advantages of Integrated Farming System

In an integrated farming system, increased production and productivity with optimal resource utilisation, recycling of waste and family labour management have been advantageous (Sasikala *et al.*, 2015). Profitability can be significantly increased since intermediate farm products like manure, draught power and crop residues are effectively utilised in IFS (Ngambeki *et al.*, 1992). Compared to the conventional mode of farming, the integrated farming system is more profitable due to the interaction of various enterprises. (Singh *et al.*, 1993). Under IFS, self-sustainability is advocated with reduced dependency on external inputs such as seeds/fertilisers etc. The comparative analysis of different modes of farming also indicates that the integration of resource management can be productive, profitable and manageable if provisions for labour and secure tenure are available (Ashby, 2001). There has been an increase in productivity due to the integration of diverse enterprises of different economic practices. Dependence on external high inputs is reduced due to agricultural waste being built into the system. The production of year-round products such as eggs, mushrooms, honey etc. enables resource-poor farmers to get out of the clutches of moneylenders. The reduction of chemical fertiliser usage is another benefit of nutrient recycling. Additionally, biogas production provides a reliable source of fuel for households. Our nation has implemented various successful Integrated Farming System (IFS) models across different regions, resulting in increased income and sustainability for small-scale farmers. Consequently, IFS plays a significant role in addressing the energy crisis.

Impact of Integrated Farming Systems

1. Economic Impact

a. **Yield** – Integration of the farm enterprises including agriculture, horticulture, poultry, piggery, fishery, apiculture, vermicomposting, etc. in IFS results in producing more yield. Higher production of rice (69.5%), vegetables (50%), fruits (65%), meat (85%), eggs (79%), and fish (57%) were observed under IFS in comparison with the conventional farming system (Ansari *et al.*, 2013). Similarly, in a study conducted by Ray *et al.* (2019), the yield improvement in IFS was recorded at 9.9-29.8% in a variety of crops and 9.92% in fruit yield compared to that of conventional shifting cultivation systems. Additionally, it was discovered that 0.04 ha of sweet potato plantations could also supplement about 12% of pigs' feed requirements while 0.1 ha of cassava plantations reduced feed requirements by an average of 27.3%. A typical yield of 1.24 t of valuable mature compost, or 53.47 kg of NPK nutrients, was also produced by the installation of vermicomposting units in IFS, allowing for the annual recycling of 3.17 t of biomass.

b. **Income** – The Integrated Farming System is a comprehensive agricultural method that aims to enhance the earnings of small and marginal farmers through a significant increase in productivity. This is accomplished by integrating various farming activities and optimizing the utilisation of crop residues and by-products from interconnected components within the farm. IFS was found to generate higher gross returns, net returns, and B:C ratios than CFS due to increased production of economic products in the corresponding IIFS components (Ansari *et al.*, 2013). The integrated farming system outperformed the traditional shifting cultivation system in terms of the annual net returns to the tribal jhumias by 471% (Ray *et al.*, 2019). Additionally, it contributes to raising the standard of living for marginal and small farmers. It enables farmers to compete in the international market and be self-sufficient. The implementation of the Integrated Farming System (IFS) provides a consistent source of employment and income for farmers by involving them in various enterprises. An example of this is the combination of oilseed cultivation with apiculture, which significantly boosts oilseed productivity by 2-3 times. Moreover, farmers can also obtain additional products like wax, pollen, bee venom, and honey, leading to increased profits. In a study conducted by Kaur *et al.* (2021), the per capita income and per acre income of the farmers are found to be higher in IFS compared to that of the conventional farming system thereby generating more annual income (Table 2). The cost of cultivation is decreased due to lesser reliance on external resources such as fertilisers, pesticides, etc. Reduction in production costs with higher net return minimises the risks and improves the overall income with efficient farm production.

Table 1. Income levels of different categories of farmers per household in Punjab (in ₹ per annum) (Kaur *et al.*, 2021).

Sl. No.	Income Source	Integrated Farming System (IFS)			Conventional Farming System (CFS)		
		Marginal Farmers	Small Farmers	All Farmers	Marginal Farmers	Small Farmers	All Farmers
1.	Farm business income	178655 (36.82)	296880 (46.04)	250247 (43.01)	154964 (48.57)	271460 (61.98)	213212 (56.33)
2.	Rent from leased out land	0 (0.00)	9357.80 (1.45)	5666.67 (0.97)	3055.56 (0.96)	4777.78 (1.09)	3916.67 (1.04)
3.	Hiring out labour in agricultural sector	1225.35 (0.25)	0 (0.00)	483.33 (0.08)	3933.33 (1.23)	666.67 (0.15)	2300 (0.61)
4.	Hiring out agricultural equipment	9901.41 (2.04)	8495.41 (1.32)	9050 (1.56)	2400 (0.75)	10444.44 (2.39)	6422.22 (1.70)
5.	Sale of seeds and manure	57.75 (0.01)	0 (0.00)	22.78 (0.01)	0 (0.00)	0 (0.00)	0 (0.00)
6.	Cattle rearing/ dairy farmer/ trade of animals	171504 (35.35)	169639 (26.31)	170374 (29.28)	25153 (7.88)	30721 (7.02)	27937 (7.38)
7	Poultry	936.11 (0.19)	11146.33 (1.73)	7118.97 (1.22)	0 (0.00)	0 (0.00)	0 (0.00)
8	Apiary	16200 (3.34)	34484.59 (5.35)	27272.33 (4.69)	0 (0.00)	0 (0.00)	0 (0.00)
9	Goat	15318.59 (3.16)	5579.08 (0.87)	9420.78 (1.62)	0 (0.00)	0 (0.00)	0 (0.00)
10.	Sheep	9008.45 (1.86)	0 (0.00)	3553.33 (0.61)	0 (0.00)	0 (0.00)	0 (0.00)
11.	Piggery	995.21 (0.21)	7734.13 (1.20)	5076 (0.87)	0 (0.00)	0 (0.00)	0 (0.00)
12.	Fishery	8711.27 (1.800	11329.36 (1.76)	10296.67 (1.77)	0 (0.00)	0 (0.00)	0 (0.00)
13.	Horticulture	22249.86 (4.59)	23400.73 (3.63)	22946.78 (3.94)	0 (0.00)	0 (0.00)	0 (0.00)
14.	Biogas plant	2642.25 (0.54)	0 (0.00)	1042.22 (0.18)	0 (0.00)	0 (0.00)	0 (0.00)
15.	Other allied activities	7312.68 (1.51)	0 (0.00)	2884.44 (0.50)	0 (0.00)	0 (0.00)	0 (0.00)
16.	Govt. services	9295.77 (1.92)	18341.28 (2.84)	14773.33 (2.54)	38266.67 (11.99)	59066.67 (13.49)	48666.76 (12.86)
17.	Private services	5915.49 (1.22)	13981.65 (2.17)	10800 (1.86)	36600 (11.47)	26146.67 (5.97)	31373.33 (8.28)
18.	Pensions and remittances	10183.10 (2.09)	21155.96 (3.28)	16827.78 (2.89)	12264.44 (3.84)	9455.56 (2.16)	10860 (2.87)
19.	Hiring out labour in non-agricultural sector	169.01 (0.03)	0 (0.00)	66.67 (0.01)	1244.44 (0.40)	0 (0.00)	622.22 (0.16)
20.	Outer sources	14873.24 (3.07)	13247.71 (2.05)	13888.89 (2.39)	41200 (12.91)	25222.22 (5.75)	33211.11 (8.77)
	Total income	485154.54 (100)	644773.03 (100)	581811.97 (1000	319081.44 (100)	437961.01 (100)	378521.22 (100)

*Figures in brackets indicates percentage

2. Social Impact

a. **Human health** – Integrated farming enterprises provide the farm family with diverse farm products including fruits, cereals, vegetables, milk, meat, and eggs which helps to improve their nutritional status and maintain a balanced diet. Given that the majority of their food came from their fields (subsistence farming), they had adequate levels of proteins and calcium in their diet. In an integrated crop-livestock farming system, the family members of smallholder farmers are provided with both grain and animal protein (Sekaran *et al.*, 2021). Additionally, this lowers medical expenses and maintains the health of the farm family.

b. **Gender** – The role of women in agriculture is important, especially in tribal and hilly regions. The profitability of the household can be improved by adopting innovative practices and utilising multiple household resources with judicious family labour. It can be achieved by empowering women with critical need-based assistance, capacity-building, and proper training in specified locations. With the proper intervention of technical, institutional, and credit support, there is scope to improve the well-being of the farmers, mainly benefiting the most vulnerable group i.e., women and children. Therefore, it is important to develop women-centric farming system models with the feminisation of agriculture (Soni *et al.*, 2014). Gender equality in agriculture will eventually improve the nutrition, health, and education outcomes for children. IFS also creates a platform for business-oriented and professional farming systems that utilize skilled youths to impart knowledge and manage intensive farming systems. Input-output supply chains for primary and secondary agriculture can be established through proper training of the youth population. The process of trans-migration can be reversed with youths from urban areas engaging in rural areas with the development of micro-business farming models that are productive, environment-friendly, profitable, and sustainable in the long term. This further helps in boosting agro-eco tourism. (Soni *et al.*, 2014).

c. **Food Security** – The Integrated Farming System encompasses multiple businesses that allow farmers to acquire a variety of high-quality products while also offering diverse sources of income. By integrating complementary IFS systems effectively, farmers can fulfil a portion or even the entirety of their dietary requirements on a small plot of land. These systems present opportunities for cultivating short-duration vegetable crops, pulses, and livestock feed, thereby enabling small and marginal farmers in India to produce the majority of their staple foods

(Paramesh *et al.*, 2022). According to Devendra and Thomas (2002), IFS is crucial for enabling poor, small, and marginal farmers to meet their protein needs through the production of eggs, milk, and meat from their livestock. Through improved resource management, the introduction of legumes, oilseed crops, vegetables, or agroforestry systems, the IFS may help to achieve food and nutritional security (Altieri *et al.*, 2012; Wezel *et al.*, 2014). Thus, IFS is essential for the expanding Indian population's future food security and nutrition.

3. Environmental Impact

a. **Soil and Nutrients** – The utilisation of organic residues from plants and animals in the Integrated Farming System reduces the reliance on external resources like agrochemicals, fertilisers, energy, and feeds. By incorporating livestock manure, the system enhances the soil's chemical and physical characteristics, thereby enriching its nutritional value through increased NPK content and other essential mineral nutrients (Kumar *et al.*, 2012). This in turn improves soil health as well as soil fertility. It also enhances the organic matter content of the soil, thus improving water infiltration and water holding capacity along with higher cation-exchange capacity. Moreover, the introduction of livestock urine and manure into the soil results in an elevation of soil pH, thereby facilitating the decomposition of organic matter and promoting microbial activity in the soil (Brouwer and Powell, 1995). Adoption of cropping techniques such as cover crops and crop rotation improves soil fertility, soil health, and water quality.

b. **Water** – In hilly agroecological situations, crop performance is primarily influenced by the distribution of annual rainfall because crops are typically grown only during the rainy or kharif season of the year and cropping intensity is close to 100%. By utilising farm ponds, Jalkunds, and other water features, farmers can effectively irrigate a diverse range of vegetables and pulses, thereby playing a crucial role in their sustenance and productivity. The provision for irrigation water in IFS promotes double cropping and as a result, cabbage and garden pea can be successfully planted during the post-rainy season. The enriched pond water can be beneficial for the irrigation of winter crops. Consequently, the IFS rainwater harvesting structures increased the average cropping intensity with the introduction of rabi crops in IFS. In a study conducted by Ansari *et al.* (2013), around 3,000-5,000 litres of rainwater are conserved as opposed to 55,000 litres in IIFS. IIFS conserves water by

1200% more than CFS in farm ponds, Jalkunds, and other similar water sources.

c. **GHGs Emission** – As a result of higher diversification of farm by-products and less dependence on external inputs in the IFS system, the smallholders on mixed crop-livestock farms are found to be more resilient to climate change and crop failure. The boundary plantation of perennial trees or fruit crops allows higher carbon sequestration and decreases the GHG emissions which counteract the negative effects of climate change. Salton *et al.* (2014) found that net GHG emissions were positive in the conventional system and negative in the IFS system. This trend was primarily caused by higher soil carbon sequestration in the IFS system, which balanced out N2O emissions. An integrated farming system with a higher number of complementary enterprises has lower greenhouse gas emissions (Rathore *et al.*, 2022). Hence, the increase in economic output by reducing the impact of greenhouse gas (GHG) emissions and lowering the impact of energy use are the main aims of IFS (Cicek *et al.*, 2011). The amount of CH_4 absorbed by IFS in Brazil was 68% less than that of traditional mono-cropping systems (Paramesh *et al.*, 2022), and according to Chen *et al.* (2011), the amount of CH_4 absorbed by IFS in temperate plains was 30% less. Management techniques like nutrient management using crop residues and compost, legumes for nitrogen fixation, and cultivation techniques like direct seeded rice/SRI method of rice cultivation in the IFS system can increase the resilience of crops to changing climates and also reduce GHG emissions. Therefore, IFS can be promoted as a viable approach to boost agricultural output and restore deteriorated pastures while reducing GHG emissions.

d. **Biodiversity** – The adoption of natural inputs and nutrient recycling techniques in the system leads to a decrease in environmental harm and an increase in farm productivity, resulting in lower cultivation costs. IFS provides various ecosystem services from agriculture like pollination, nitrogen fixation, nutrient recycling, improved soil quality, water penetration, and reduced economic losses due to crop failure through different cropping systems such as intercropping, mixed cropping, sequential cropping, etc. Additionally, it promotes soil microbial diversity by incorporating compost, manure, or duck droppings in rice-fish-duck cultures. It also promotes the coexistence of a variety of species in a local area, which adds complexity to the food web, thereby preserving the biodiversity of the region. IFS also encourages non-crop biodiversity, first by using ecological management techniques like using insecticides and small-scale fertilisers, and then by using organic management techniques. Additionally, incorporating timber trees and fruit trees into the agricultural system not only boosts biodiversity, but also acts as a protective barrier against wind and enhances the visual appeal of the farm. Moreover, this integration provides a source of income and ensures nutritional stability for the farming household (Sunderland, 2011). Thus, it restores biodiversity while increasing the whole system's economic and agronomic productivity.

Challenges in Integrated Farming System

The challenge of implementing integrated farming systems is increasing due to the decrease in land availability. It is crucial to optimise and integrate different agricultural enterprises while ensuring access to credit for the effective design, development, and validation of integrated farming systems. In order to envision a successful IFS model, technical empowerment through proper provisioning of suitable training and demonstration is needed. Due to the low volume of products, costs of transport and labour input, selling of produce by the marginal and small farmers is another serious constraint. Hence, the establishment of a market chain for the perennial marketing of the farm products at the farm gate is an important issue that needs to be addressed in integrated farming systems (Manjunatha, 2014).

Future Thrust

- Scientific validation and proper documentation of the ITKs available in the farming community.
- Modules developed under integrated farming system need refinement by adding profitable components.
- Development of an IFS model that is both environmentally sustainable and economically viable for various landholding sizes is crucial.
- Establishment of a comprehensive framework for integrated farming systems in diverse agro-climatic conditions necessitates a precise definition of sustainable practices.
- The identification of constraints in the implementation of the integrated farming system within a specific area or locality is crucial
- Thorough assessment of the nutrient dynamics of the soil by taking into account the cropping technique and the proper recycling of organic resources.

Conclusion

IFS has the potential to enhance soil health and productivity by incorporating organic residues such as animal and plant wastes over time, resulting in fewer environmental hazards and increased profit margins. In addition to these benefits, IFS also offers advantages such as reduced cultivation costs, increased employment opportunities, and self-sufficiency in nutritional requirements. However, the successful implementation of IFS models in different ecological systems requires participatory trials and interventions at multiple levels, specifically on farmers' fields. Undoubtedly, IFS contributes to improved profitability, livelihoods, employment generation, and natural resource conservation by reducing production costs and minimising risks. Therefore, the IFS approach holds promise in efficiently conserving the resource base through the recycling of residues and wastes within the system.

References

Altieri, M. A., Funes- Monzote, F. R. and Petersen, P. (2012). Agroecologically efficient agricultural systems for smallholder farmers: contributions to food sovereignty. Agronomy for Sustainable Development, 32(1), 1– 13.

Ansari, Meraj, Prakash, Namarata, Baishya, Lohit, Punitha, Periaswamy Yadav, J., Sharma, P., Sailo, Blessa and Ansari, M.H. (2013). Comparative study on conventional and improved integrated farming systems for sustainable production, income generation and employment opportunity among the tribal farmers in hilly Regions of Manipur. Indian Journal of Agricultural Sciences. 83. 765–72.

Ashby, J.A. (2001). Integrating research on food and the environment: An exit strategy from the rational fool syndrome in agricultural science. Ecol. Soc, 5.

Ayyappan, S., Arunachalam, A., (2014). Crop diversification for social security of farmer in India. In: Rana, D.S., Singh, A.K. (Eds.), Souvenir. Indian Society of Agronomy, New Delhi, pp. 1–6.

Brouwer J and Powell J M. (1995). Soil aspects of nutrient cycling in a manure experiment in Niger. Livestock and Sustainable Nutrient Cycling in Mixed Farming Systems of Sub-Saharan Africa. Vol II: Technical papers. Powel J M, Fernandez-Rivera S, Williams T O and Renard C (Eds). Proceedings of an International Conference held during 22-26 November 1993 in Addis Ababa, Ethiopia, pp 349-60.

Chen, W., Wolf, B., Zheng, X., Yao, Z., Butterbach- Bahl, K., Brüggemann, N., Liu, C., Han, S., & Han, X. (2011). Annual methane uptake by temperate semiarid steppes as regulated by stocking rates, aboveground plant biomass and topsoil air permeability. Global Change Biology, 17(9), 2803– 2816.

Devendra, C., & Thomas, D. (2002). Smallholder farming systems in Asia. Agricultural Systems, 71(1– 2), 17– 25.

Edwards P, Pullin R S V and Gartner J A. (1988). Research and education for the development of integrated crop-livestock-fish farming systems in the tropics. ICLARM Studies and Reviews 16, Manilla, Philippines.

FAO, (2010). Sustainable crop production intensification through an ecosystem approach and an enabling environment: capturing efficiency through ecosystem services and management, FAO Committee on Agriculture, June 16-19.

Gill, M.S., Singh, J.P., Gangwar, K.S., 2010. Integrated farming systems and agriculture sustainability. Indian J. Agron. 54(20), 128–139.

Gill M.S., Samra J.S., and Singh G., (2005). Integrated farming system for realizing high productivity under shallow water-table conditions. Research bulletins, Department of Agronomy, PAU, Ludhiana, pp. 1-29.

Gill M S, Singh J P and Gangwar K S. 2009. Integrated farming System and agriculture sustainability. Indian Journal of Agronomy. 54(2): 128–39.

Gupta V, Rai P K and Risam K S. (2012). Integrated crop-livestock farming systems: A strategy for resource conservation and environmental sustainability. Indian Research Journal of Extension Education, 2: 49–54.

IAASTD, (2009). Agriculture at the Crossroads, International Assessment of Agricultural Knowledge, Science and Technology for Development, Washington, DC, Island Press.

Kumar, S., Shivani, Singh, S.S., 2013. Sustainable food and nutritional security through IFS. Indian Farm. 63 (7), 30–36.

Kumar, S., Singh S. S, Kumar U., Dey, A. and Shivani. 2013. Sustainability of integrated farming systems in the eastern region. Natural Resource Conservation – Emerging Issues and Future Challenges, pp. 455-64.

Kumar S. Integrated Farming System models for food and nutritional security. Lecture Delivered During Model Training Course on Gender Perspective in Integrated Farming system w.e.f January. at ICAR Research Complex for Eastern Region, Patna, Bihar, 2013, 17-24.

Lal, R. and Miller F P. 1990. Sustainable farming for tropics. Sustainable Agriculture: Issues and Prospective, Vol 1 pp 69-89. Singh R P (Ed). Indian Society of Agronomy, IARI, New Delhi.

Maheswarappa, H.P., Hegde, M.R., Dhanpal, R., Biddappa, C.C., 1998. Mixed farming in coconut garden: its impact on soil, physical, chemical properties, coconut nutrition and yield. J. Plant. Crops. 26 (2), 139–143.

Maheswarappa, H.P., Hegde, M.R., Dhanpal, R., Sairam, C.V., Singh, T.V., 2001. Impact of integrated mixed farming systems in coconut garden on coconut yield and economic analysis. Indian J. Agron. 46 (1), 56–63.

Mali Hansram, Kumar Amit and Katara Pawan. 2014. Integrated farming system for irrigated and rainfed conditions. (In) Proceedings of National Symposium on Agricultural Diversification for Sustainable Livelihood and Environmental Security, held during 18-20 November 2014 at Ludhiana, Punjab, p 546.

Manjunath, B.L., Itnal, C.J., 2003. Integrated farming system in enhancing the productivity of marginal rice (Oryza sativa) holdings in Goa. Indian J. Agron. 48 (1), 1–3.

Manjunatha, S.B., Shivmurthy, D., Sunil, A.S., Nagaraj, M.V. and Basavesha K.N. (2014). Integrated Farming System – A Holistic Approach: A Review. Research and Reviews: Journal of Agriculture and Allied Sciences. e-ISSN: 2319-9857 p-ISSN: 2347-226X.

Manjunath B L and Singh M P. 2012. Farming system options for different topographical situations of west coast region. Extended Summaries Vol 3: 3rd International Agronomy Congress held during 26-30 November 2012 at New Delhi, pp 988-90.

Mohanty, D., Patnaik, S.C., Jeevan Das, P., Parida, N.K. and Nedunchezhiyan, M. (2010). Sustainable livelihood: a success story of a tribal farmer. Orissa Review, September: pp. 41 – 43.

Ngambeki, D.S., Deuson, R.R. and Preckel, P.V. (1992). Integrating livestock into farming systems in northern Cameroon. Agricultural Systems, 38: pp. 319-338.

Okigbo, B.N., (1995). Major farming systems of the lowland savanna of SSA and the potential for improvement. In: Proceedings of the IITA/FAO workshop, Ibadan, Nigeria.

Pandey, P. R., Gupta, J. K., Narvariya, R. K., Meena, S. C., & Narwariya, D. (2019). Constraints faced by farmers in adoption of integrated farming system in Vindhyan Plateau of Madhya Pradesh. Plant Archives, 19(2), 512–514.

Paramesh, V., Ravisankar, N., Behera, U., Arunachalam, V., Kumar, P., Solomon Rajkumar, R., Dhar Misra, S., Mohan Kumar, R., Prusty, A. K., Jacob, D., Panwar, A. S., Mayenkar, T., Reddy, V. K., & Rajkumar, S. (2022). Integrated farming system approaches to achieve food and nutritional security for enhancing profitability, employment, and climate resilience in India. Food and Energy Security, 11, e321. https://doi. org/10.1002/fes3.321

Pathak, P. and Laryea K.B. (1995). Soil and water conservation in the Indian SAT; principles and improved practices. In: Sustainable development of dryland agriculture in India. (Ed.) R.P. Singh. Scientific Publishers, Jodhpur. pp. 83-92.

Patra, A.K. 2013. Agroforestry: Theory & Practices. New India Publishing Agency, New Delhi. 242 pp.

Ponnusamy, K and Devi, K. (2017). Impact of Integrated Farming System Approach on Doubling Farmers' Income. Agricultural Economics Research Review. 30: 233-240.

Pushpa, J. (2010). Constraints in various integrated farming systems. Agriculture Update, 5(3/4), 370–374.

Rangasamy, A., Venkitasamy, R., Jayanthi, C., Purushothaman, S. and Palaniappan, S.P. (1995) Rice based farming system: A viable approach. Indian Farming, 46 (4): 27-29.

Rao, K.T., Rao M.V.S. and Patro T.S.S.K (2019). Integrated Farming Systems: A potential Tool for Doubling Farmer's Income. International Journal of Current Microbiology and Applied Sciences. 8(9): 2629- 2642.

Ravisankar, N., Subrmani, T., Ambast, S.K., Srivasyava, R.C., 2010. Enhancing farm income in an island ecosystem. Indian Farm. 60 (6), 16–19.

Ray, Sanjay & Chatterjee, Dibyendu & Rajkhowa, D.J. & Baishya, Santosh & Hazarika, Samarendra & Paul, Sudipta. (2020). Effects of integrated farming system and rainwater harvesting on livelihood improvement in North-Eastern region of India compared to traditional shifting cultivation: evidence from action research. Agroforestry Systems. 94. 10.1007/s10457-019-00406-3.

Salton, J. C., Mercante, F. M., Tomazi, M., Zanatta, J. A., Concenço, G., Silva, W. M., & Retore, M. (2014). Integrated crop-livestock system in tropical Brazil: Toward a sustainable production system. Agriculture, Ecosystems & Environment, 190, 70– 79.

Sasikala V., Tiwari R., Saravanam M. (2015). A review on integrated farming system journal of international academic research for multidisciplinary ISSN: 2320-5083, Volume 3, Issue 7.

Sekaran, U., Lai, L., Ussiri, D. A., Kumar, S., & Clay, S. (2021). Role of integrated crop-livestock systems in improving agriculture production and addressing food security–A review. Journal of Agriculture and Food Research, 5, 100190.

Shukla, A.K., Shukla, N.D., Singh, V.K., 2002. Farming system approach for food security

Singh, K.P., Singh S.N., Kumar H., Kadian V.S. and Saxena K.K. (1993). Economic analysis of different farming systems followed on small and marginal land holdings in Haryana. Haryana Journal of Agronomy, 9: pp. 122-125.

Soni, Rajju & Katoch, Mittu & Ladohia, Rajesh. (2014). Integrated Farming Systems - A Review. 7. 36-42.

Sunderland, T. C. H. (2011). Food security: Why is biodiversity important? International Forestry Review, 13(3), 265– 274.

Tripathi, S.C., and Rathi, R.C. (2011). Livestock farming system module for hills. In: Souvenir. National symposium on technological interventions for sustainable agriculture, 3rd - 5th May, GBPUAT, hill campus, Ranichuri, pp. 103 -104.

Velu, V. (2011). Soil fertility management strategies through INM in rainfed agriculture. In: CAFTA training manual on Productivity enhancement in drylands through diversified farming options, TNAU, Coimbatore, Jan. 27 – Feb.16. pp. 81-92

Vincke M. 1988. La rizipisciculture et les elevageassociesenAfrique. CIFA Technical Paper (4) Suppl 1: 659–707.

Wezel, A., Casagrande, M., Celette, F., Vian, J.- F., Ferrer, A., & Peigné, J. (2014). Agroecological practices for sustainable agriculture. A Review. Agronomy for Sustainable Development, 34(1), 1–20.

Yassi, A., Kaimuddin and Ekawati, I. (2020). 2020 IOP Conf. Ser.: Earth Environ. Sci. 575.

Index